CAMBRIDGE LIBR.
Books of enduring

Philosophy

This series contains both philosophical texts and critical essays about
philosophy, concentrating especially on works originally published in the
eighteenth and nineteenth centuries. It covers a broad range of topics including
ethics, logic, metaphysics, aesthetics, utilitarianism, positivism, scientific
method and political thought. It also includes biographies and accounts of
the history of philosophy, as well as collections of papers by leading figures.
In addition to this series, primary texts by ancient philosophers, and works
with particular relevance to philosophy of science, politics or theology, may be
found elsewhere in the Cambridge Library Collection.

Collected Essays

Known as 'Darwin's Bulldog', the biologist Thomas Henry Huxley (1825–95)
was a tireless supporter of the evolutionary theories of his friend Charles
Darwin. Huxley also made his own significant scientific contributions, and
he was influential in the development of science education despite having had
only two years of formal schooling. He established his scientific reputation
through experiments on aquatic life carried out during a voyage to Australia
while working as an assistant surgeon in the Royal Navy; ultimately he
became President of the Royal Society (1883–5). Throughout his life Huxley
struggled with issues of faith, and he coined the term 'agnostic' to describe his
beliefs. This nine-volume collection of Huxley's essays, which he edited and
published in 1893–4, demonstrates the wide range of his intellectual interests.
In Volume 6, Huxley focuses on the philosopher David Hume (1711–76),
discussing his life and his philosophical and intellectual work.

Cambridge University Press has long been a pioneer in the reissuing of out-of-print titles from its own backlist, producing digital reprints of books that are still sought after by scholars and students but could not be reprinted economically using traditional technology. The Cambridge Library Collection extends this activity to a wider range of books which are still of importance to researchers and professionals, either for the source material they contain, or as landmarks in the history of their academic discipline.

Drawing from the world-renowned collections in the Cambridge University Library, and guided by the advice of experts in each subject area, Cambridge University Press is using state-of-the-art scanning machines in its own Printing House to capture the content of each book selected for inclusion. The files are processed to give a consistently clear, crisp image, and the books finished to the high quality standard for which the Press is recognised around the world. The latest print-on-demand technology ensures that the books will remain available indefinitely, and that orders for single or multiple copies can quickly be supplied.

The Cambridge Library Collection will bring back to life books of enduring scholarly value (including out-of-copyright works originally issued by other publishers) across a wide range of disciplines in the humanities and social sciences and in science and technology.

Collected Essays

VOLUME 6: HUME, WITH HELPS TO THE STUDY OF BERKELEY

THOMAS HENRY HUXLEY

CAMBRIDGE
UNIVERSITY PRESS

CAMBRIDGE UNIVERSITY PRESS

Cambridge, New York, Melbourne, Madrid, Cape Town,
Singapore, São Paolo, Delhi, Tokyo, Mexico City

Published in the United States of America by Cambridge University Press, New York

www.cambridge.org
Information on this title: www.cambridge.org/9781108040563

© in this compilation Cambridge University Press 2011

This edition first published 1894
This digitally printed version 2011

ISBN 978-1-108-04056-3 Paperback

This book reproduces the text of the original edition. The content and language reflect
the beliefs, practices and terminology of their time, and have not been updated.

Cambridge University Press wishes to make clear that the book, unless originally published
by Cambridge, is not being republished by, in association or collaboration with, or
with the endorsement or approval of, the original publisher or its successors in title.

COLLECTED ESSAYS

By T. H. HUXLEY

VOLUME VI

HUME:

WITH HELPS TO THE STUDY OF

BERKELEY

ESSAYS

BY

THOMAS H. HUXLEY

London
MACMILLAN AND CO.
1894

RICHARD CLAY AND SONS, LIMITED,
LONDON AND BUNGAY.

PREFACE

In two essays upon the life and work of Descartes, which will be found in the first volume of this collection, I have given some reasons for my conviction that he, if any one, has a claim to the title of father of modern philosophy. By this I mean that his general scheme of things, his conceptions of scientific method and of the conditions and limits of certainty, are far more essentially and characteristically modern than those of any of his immediate predecessors and successors. Indeed, the adepts in some branches of science had not fully mastered the import of his ideas so late as the beginning of this century.

The conditions of this remarkable position in the world of thought are to be found, as usual, primarily, in motherwit, secondarily, in circumstance. Trained by the best educators of the seventeenth century, the Jesuits; naturally endowed with a dialectic grasp and subtlety, which even they could hardly improve; and with a passion for getting at the truth, which even they could

hardly impair, Descartes possessed, in addition,
a rare mastery of the art of literary expression.
If the "Discours de la Méthode" had no other
merits, it would be worth study for the sake of
the luminous simplicity and sincerity of its style.

A mathematician of the very first rank,
Descartes knew all that was to be known of
mechanical and optical science in his day; he was
a skilled and zealous practical anatomist; he was
one of the first to recognise the prodigious importance of the discovery of his contemporary
Harvey; and he penetrated more deeply into
the physiology of the nervous system than any
specialist in that science, for a century, or more,
after his time. To this encyclopædic and yet
first-hand acquaintance with the nature of things,
he added an acquaintance with the nature of
men (which is a much more valuable chapter of
experience to philosophers than is commonly
imagined), gathered in the opening campaigns of
the Thirty Years' War, in wide travels, and amidst
that brilliant French society in which Pascal was
his worthy peer. Even a "Traité des Passions," to
be worth anything, must be based upon observation
and experiment; and, in this subject, facilities for
laboratory practice of the most varied and extensive character were offered by the Paris of
Mazarin and the Duchesses; the Paris, in which
Descartes' great friend and ally, Father Mersenne,
reckoned atheists by the thousand; and, in which,

political life touched the lowest depths of degra-
dation, amidst the chaotic personal intrigues of
the Fronde. Thus endowed, thus nurtured, thus
tempered in the fires of experience, it is intelligible enough that a resolute, clear-headed man,
haunted from his youth up, as he tells us, with
an extreme desire to learn how to distinguish
truth from falsehood, in order to see his way
clearly and walk surely through life,[1] should have
early come to the conclusion, that the first thing to
be done was to cast aside, at any rate temporarily,
the crutches of traditional, or other, authority; and
stand upright on his own feet, trusting to no
support but that of the solid ground of fact.

It was in 1619, while meditating in solitary
winter quarters, that Descartes (being about the
same age as Hume when he wrote the "Treatise on
Human Nature") made that famous resolution, to
"take nothing for truth without clear knowledge
that it is such," the great practical effect of which
is the sanctification of doubt; the recognition that
the profession of belief in propositions, of the truth
of which there is no sufficient evidence, is immoral;
the discrowning of authority as such; the repudi-
ation of the confusion, beloved of sophists of all
sorts, between free assent and mere piously gagged
dissent; and the admission of the obligation to
reconsider even one's axioms on due demand.

These, if I mistake not, are the notes of the

[1] *Discours de la Méthode.* 1ᵉ Partie.

modern, as contrasted with the ancient spirit.
It is true that the isolated greatness of Socrates
was founded on intellectual and moral character-
istics of the same order. He also persisted in
demanding that no man should "take anything
for truth without a clear knowledge that it is
such," and so constantly and systematically shocked
authority and shook traditional security, that the
fact of his being allowed to live for seventy years,
if one comes to think of it, is evidence of the
patient and tolerant disposition of his Athenian
compatriots, which should obliterate the memory of
the final hemlock. That which it may be well for
us not to forget is, that the first-recorded judicial
murder of a scientific thinker was compassed and
effected, not by a despot, nor by priests, but was
brought about by eloquent demagogues, to whom,
of all men, thorough searchings of the intellect
are most dangerous and therefore most hateful.

The first agnostic, the man who, so far as
the records of history go, was the first to see that
clear knowledge of what one does not know
is just as important as knowing what one does
know, had no true disciples; and the greatest of
those who listened to him, if he preserved the
fame of his master for all time, did his best to
counteract the impulse towards intellectual clearness which Socrates gave. The Platonic philosophy is probably the grandest example of the
unscientific use of the imagination extant; and it

would be hard to estimate the amount of detriment to clear thinking effected, directly and indirectly, by the theory of ideas, on the one hand, and by the unfortunate doctrine of the baseness of matter, on the other.

Ancient thought, so far as it is positive, fails on account of its neglect to criticise its assumptions; so far as it is negative, it fails, because it forgets that proof of the inconsistencies of the terms in which we symbolise things has nothing to do with the cogency of the logic of facts. The negations of Pyrrhonism are as shallow, as the assumptions of Platonism are empty. Modern thought has by no means escaped from perversions of the same order. But, thanks to the sharp discipline of physical science, it is more and more freeing itself from them. In face of the incessant verification of deductive reasoning by experiment, Pyrrhonism has become ridiculous; in face of the ignominious fate which always befalls those who attempt to get at the secrets of nature, or the rules of conduct, by the high *a priori* road, Platonism and its modern progeny show themselves to be, at best, splendid follies.

The development of exact natural knowledge in all its vast range, from physics to history and criticism, is the consequence of the working out, in this province, of the resolution to " take nothing for truth without clear knowledge that it is such ; " to consider all beliefs open to criticism; to regard the value of authority as neither greater nor less, than as much as it can prove itself to be worth. The modern spirit is not the spirit "which always denies," delighting only in destruction ; still less is it that which builds castles in the air rather than not construct; it is that spirit which works and will work "without haste and without rest," gathering harvest after harvest of truth into its barns and devouring error with unquenchable fire.

In the reform of philosophy, since Descartes, I think that the greatest and the most fruitful results of the activity of the modern spirit—it may be, the only great and lasting results—are those first presented in the works of Berkeley and of Hume.

The one carried out to its logical result the Cartesian principle, that absolute certainty attaches only to the knowledge of facts of consciousness; the other, extended the Cartesian criticism to the whole range of propositions commonly "taken for truth ;" proved that, in a multitude of important instances, so far from possessing "clear knowledge" that they may be so taken, we have none at all; and that our duty therefore is to remain silent; or to express, at most, suspended judgment.

My earliest lesson on this topic was received from Hume's keen-witted countryman Hamilton;

afterwards I learned it, more fully, from the fountain head, the "Discours de la Méthode"; then from Berkeley and from Hume themselves. So that when, in 1878, my friend Mr. John Morley asked me to write an account of Hume for the "English Men of Letters" series, I thought I might undertake the business, without too much presumption; also, with some hope of passing on to others the benefits which I had received from the study of Hume's works. And, however imperfect the attempt may be, I have reason to believe that it has fulfilled its purpose. I hoped, at one time, to be able to add an analogous exposition of Berkeley's views; and, indeed, undertook to supply it. But the burdens and distractions of a busy life led to the postponement of this, as of many other projects, till too late. My statement of Hume's philosophy will have to be provided with its counterpart and antithesis by other hands. But I have appended to the "Hume" a couple of preliminary studies, which may be of use to students of Berkeley.

One word, by way of parting advice to the rising generation of English readers. If it is your desire to discourse fluently and learnedly about philosophical questions, begin with the Ionians and work steadily through to the latest new speculative treatise. If you have a good memory and a fair knowledge of Greek, Latin, French, and

German, three or four years spent in this way should enable you to attain your object.

If, on the contrary, you are animated by the much rarer desire for real knowledge; if you want to get a clear conception of the deepest problems set before the intellect of man, there is no need, so far as I can see, for you to go beyond the limits of the English tongue. Indeed, if you are pressed for time, three English authors will suffice; namely, Berkeley, Hume, and Hobbes.

If you will lay your minds alongside the works of these great writers—not with the view of merely ascertaining their opinions, still less for the purpose of indolently resting on their authority, but to the end of seeing for yourselves how far what each says has its foundation in right reason—you will have had as much sound philosophical training as is good for any one but an expert. And you will have had the further advantage of becoming familiar with the manner in which three of the greatest masters of the English language have handled that noble instrument of thought.

T. H. HUXLEY.

HODESLEA, EASTBOURNE,
January, 1894.

CONTENTS

HUME

PART I.—HUME'S LIFE.

I

PART II.—HUME'S PHILOSOPHY.

I

HELPS TO THE STUDY OF BERKELEY

PART I

HUME'S LIFE

HUME

CHAPTER I

EARLY LIFE: LITERARY AND POLITICAL
WRITINGS

DAVID HUME was born in Edinburgh on the 26th
of April (O.S.), 1711. His parents were then
residing in the parish of the Tron church,
apparently on a visit to the Scottish capital, as
the small estate which his father, Joseph Hume,
or Home, inherited, lay in Berwickshire, on the
banks of the Whitadder or Whitewater, a few
miles from the border, and within sight of English
ground. The paternal mansion was little more
than a very modest farmhouse,[1] and the property
derived its name of Ninewells from a considerable

---

[1] A picture of the house, taken from Drummond's *History of
Noble British Families*, is to be seen in Chambers's *Book of Days*
(April 26th) ; and if, as Drummond says, "It is a favourable
specimen of the best Scotch lairds' houses," all that can be said
is that the worst Scotch lairds must have been poorly lodged
indeed.

spring, which breaks out on the slope in front of the house, and falls into the Whitadder.

Both mother and father came of good Scottish families—the paternal line running back to Lord Home of Douglas, who went over to France with the Douglas during the French wars of Henry V. and VI. and was killed at the battle of Verneuil. Joseph Hume died when David was an infant, leaving himself and two elder children, a brother and a sister, to the care of their mother, who is described by David Hume in "My Own Life" as "a woman of singular merit, who though young and handsome devoted herself entirely to the rearing and education of her children." Mr. Burton says: "Her portrait, which I have seen, represents a thin but pleasing countenance, expressive of great intellectual acuteness;" and as Hume told Dr. Black that she had "precisely the same constitution with himself" and died of the disorder which proved fatal to him, it is probable that the qualities inherited from his mother had much to do with the future philosopher's eminence. It is curious, however, that her estimate of her son in her only recorded, and perhaps slightly apocryphal utterance, is of a somewhat unexpected character. "Our Davie's a fine good-natured crater, but uncommon wake-minded." The first part of the judgment was indeed verified by "Davie's" whole life; but one might seek in vain for signs of what is commonly understood as

"weakness of mind" in a man who not only
showed himself to be an intellectual athlete, but
who had an eminent share of practical wisdom
and tenacity of purpose. One would like to know,
however, when it was that Mrs. Hume committed
herself to this not too flattering judgment of her
younger son. For as Hume reached the mature
age of four and thirty, before he obtained any
employment of sufficient importance to convert
the meagre pittance of a middling laird's younger
brother into a decent maintenance, it is not improbable that a shrewd Scots wife may have
thought his devotion to philosophy and poverty to
be due to mere infirmity of purpose. But she
lived till 1749, long enough to see more than the
dawn of her son's literary fame and official importance, and probably changed her mind about
" Davie's " force of character.

David Hume appears to have owed little to
schools or universities. There is some evidence
that he entered the Greek class in the University
of Edinburgh in 1723—when he was a boy of
twelve years of age—but it is not known how long
his studies were continued, and he did not gradu-
ate. In 1727, at any rate, he was living at
Ninewells, and already possessed by that love of
learning and thirst for literary fame, which, as
" My Own Life " tells us, was the ruling passion
of his life and the chief source of his enjoyments.
A letter of this date, addressed to his friend

Michael Ramsay, is certainly a most singular production for a boy of sixteen. After sundry quotations from Virgil the letter proceeds :—

"The perfectly wise man that outbraves fortune, is much greater than the husbandman who slips by her ; and, indeed, this pastoral and saturnian happiness I have in a great measure come at just now. I live like a king, pretty much by myself, neither full of action nor perturbation—*molles somnos.* This state, however, I can foresee is not to be relied on. My peace of mind is not sufficiently confirmed by philosophy to with-stand the blows of fortune. This greatness and elevation of soul is to be found only in study and contemplation. This alone can teach us to look down on human accidents. You must allow [me] to talk thus like a philosopher : 'tis a subject I think much on, and could talk all day long of."

If David talked in this strain to his mother her tongue probably gave utterance to " Bless the bairn ! " and, in her private soul, the epithet "wake-minded " may then have recorded itself. But, though few lonely, thoughtful, studious boys of sixteen give vent to their thoughts in such stately periods, it is probable that the brooding over an ideal is commoner at this age, than fathers and mothers, busy with the cares of practical life, are apt to imagine.

About a year later, Hume's family tried to launch him into the profession of the law; but, as he tells us, " while they fancied I was poring upon Voet and Vinnius, Cicero and Virgil were the authors which I was secretly devouring," and the attempt seems to have come to an abrupt termin-

ation. Nevertheless, as a very competent authority[1] wisely remarks :—

"There appear to have been in Hume all the elements of
which a good lawyer is made : clearness of judgment, power of
rapidly acquiring knowledge, untiring industry, and dialectic
skill : and if his mind had not been preoccupied, he might have
fallen into the gulf in which many of the world's greatest
geniuses lie buried—professional eminence ; and might have
left behind him a reputation limited to the traditional recollec-
tions of the Parliament house, or associated with important
decisions. He was through life an able, clear-headed man of
business, and I have seen several legal documents written in
his own hand and evidently drawn by himself. They stand
the test of general professional observation ; and their writer,
by preparing documents of facts of such a character on his own
responsibility, showed that he had considerable confidence in
his ability to adhere to the forms adequate for the occasion.
He talked of it as 'an ancient prejudice industriously propagated
by the dunces in all countries, that *a man of genius is unfit for
business,*' and he showed, in his general conduct through life,
that he did not choose to come voluntarily under this proscrip-
tion."

Six years longer Hume remained at Ninewells
before he made another attempt to embark in a
practical career—this time commerce—and with a
like result. For a few months' trial proved that
kind of life, also, to be hopelessly against the
grain.

It was while in London, on his way to Bristol,
where he proposed to commence his mercantile

[1] Mr. John Hill Burton, in his valuable *Life of Hume*, on
which, I need hardly say, I have drawn freely for the materials
of the present biographical sketch.

life, that Hume addressed to some eminent
London physician (probably, as Mr. Burton
suggests, Dr. George Cheyne) a remarkable letter.
Whether it was ever sent seems doubtful; but it
shows that philosophers as well as poets have
their Werterian crises, and it presents an interesting parallel to John Stuart Mill's record of the
corresponding period of his youth. The letter is
too long to be given in full, but a few quotations
may suffice to indicate its importance to those who
desire to comprehend the man.

"You must know then that from my earliest infancy I found
always a strong inclination to books and letters. As our
college education in Scotland, extending little further than the
languages, ends commonly when we are about fourteen or
fifteen years of age, I was after that left to my own choice
in my reading, and found it incline me almost equally to books
of reasoning and philosophy, and to poetry and the polite
authors. Every one who is acquainted either with the
philosophers or critics, knows that there is nothing yet established in either of these two sciences, and that they contain
little more than endless disputes, even in the most fundamental
articles. Upon examination of these, I found a certain boldness
of temper growing on me, which was not inclined to submit to
any authority in these subjects, but led me to seek out some
new medium, by which truth might be established. After
much study and reflection on this, at last, when I was about
eighteen years of age, there seemed to be opened up to me
a new scene of thought, which transported me beyond measure,
and made me, with an ardour natural to young men, throw up
every other pleasure or business to apply entirely to it. The
law, which was the business I designed to follow, appeared
nauseous to me, and I could think of no other way of pushing
my fortune in the world, but that of a scholar and philosopher.

I was infinitely happy in this course of life for some months ; till at last, about the beginning of September, 1729, all my ardour seemed in a moment to be extinguished, and I could no longer raise my mind to that pitch, which formerly gave me such excessive pleasure."

This " decline of soul " Hume attributes, in part, to his being smitten with the beautiful representation of virtue in the works of Cicero, Seneca, and Plutarch, and being thereby led to discipline his temper and his will along with his reason and understanding.

"I was continually fortifying myself with reflections against death, and poverty, and shame, and pain, and all the other calamities of life."

And he adds very characteristically :—

" These no doubt are exceeding useful when joined with an active life, because the occasion being presented along with the reflection, works it into the soul, and makes it take a deep impression : but, in solitude, they serve to little other purpose than to waste the spirits, the force of the mind meeting no resistance, but wasting itself in the air, like our arm when it misses its aim."

Along with all this mental perturbation, symptoms of scurvy, a disease now. almost unknown among landsmen, but which, in the days of winter salt meat, before root crops flourished in the Lothians, greatly plagued our forefathers, made their appearance. And, indeed, it may be suspected that physical conditions were, at first, at the bottom of the whole business ; for, in 1731, a ravenous appetite set in and, in six weeks from being tall, lean, and raw-boned, Hume says he became sturdy and robust, with a ruddy complexion and a cheerful countenance—eating, sleeping, and feeling well, except that the capacity for intense mental application seemed to be gone. He, therefore, determined to seek out a more active life; and, though he could not and would not "quit his pretensions to learning, but with his last breath," he resolved "to lay them aside for some time, in order the more effectually to resume them."

The careers open to a poor Scottish gentleman in those days were very few; and, as Hume's option lay between a travelling tutorship and a stool in a merchant's office, he chose the latter.

"And having got recommendation to a considerable trader in Bristol, I am just now hastening thither, with a resolution to forget myself, and everything that is past, to engage myself, as far as is possible, in that course of life, and to toss about the world from one pole to the other, till I leave this distemper behind me." [1]

But it was all of no use—Nature would have her way—and in the middle of 1736, David Hume, aged twenty-three, without a profession or any assured means of earning a guinea; and having doubtless, by his apparent vacillation, but real tenacity of purpose, once more earned the

[1] One cannot but be reminded of young Descartes' renunciation of study for soldiering.

title of "wake-minded" at home; betook himself
to a foreign country.

"I went over to France, with a view of prosecuting my
studies in a country retreat: and there I laid that plan of life
which I have steadily and successfully pursued. I resolved to
make a very rigid frugality supply my deficiency of fortune,
to maintain unimpaired my independency, and to regard every
object as contemptible except the improvement of my talents
in literature."[1]

Hume passed through Paris on his way to
Rheims, where he resided for some time; though
the greater part of his three years' stay was spent
at La Flêche, in frequent intercourse with the
Jesuits of the famous college in which Descartes
was educated. Here he composed his first work,
the "Treatise of Human Nature"; though it
would appear from the following passage in the
letter to Cheyne, that he had been accumulating
materials to that end for some years before he left
Scotland.

"I found that the moral philosophy transmitted to us by
antiquity laboured under the same inconvenience that has been
found in their natural philosophy, of being entirely hypothetical, and depending more upon invention than experience:
every one consulted his fancy in erecting schemes of virtue and
happiness, without regarding human nature, upon which every
moral conclusion must depend."

This is the key-note of the "Treatise"; of
which Hume himself says apologetically, in one of
his letters, that it was planned before he was

[1] *My Own Life.*

twenty-one and composed before he had reached the age of twenty-five.[1]

Under these circumstances, it is probably the most remarkable philosophical work, both intrinsically and in its effects upon the course of thought, that has ever been written. Berkeley, indeed, published the "Essay Towards a New Theory of Vision," the "Treatise Concerning the Principles of Human Knowledge," and the "Three Dialogues," between the ages of twenty-four and twenty-eight ; and thus comes very near to Hume, both in precocity and in influence ; but his investigations are more limited in their scope than those of his Scottish contemporary.

The first and second volumes of the "Treatise," containing Book I., "Of the Understanding," and Book II., "Of the Passions," were published in January, 1739. The publisher gave fifty pounds for the copyright; which is probably more than an unknown writer of twenty-seven years of age would get for a similar work, at the present time. But, in other respects, its success fell far short of Hume's expectations. In a letter dated the 1st of June, 1739, he writes,—

"I am not much in the humour of such compositions at present, having received news from London of the success of

---

[1] Letter to Gilbert Elliot of Minto, 1751. "So vast an undertaking, planned before I was one-and-twenty, and composed before twenty-five, must necessarily be very defective. I have repented my haste a hundred and a hundred times."

my 'Philosophy,' which is but indifferent, if I may judge by the sale of the book, and if I may believe my bookseller."

This, however, indicates a very different reception from that which Hume, looking through the inverted telescope of old age, ascribes to the "Treatise" in "My Own Life."

"Never literary attempt was more unfortunate than my 'Treatise of Human Nature.' It fell *deadborn from the press* without reaching such a distinction as even to excite a murmur among the zealots."

As a matter of fact, it was fully, and, on the whole, respectfully and appreciatively, reviewed in the "History of the Works of the Learned" for November, 1739.[1] Whoever the reviewer may have been, he was a man of discernment, for he says that the work bears "incontestable marks of a great capacity, of a soaring genius, but young, and not yet thoroughly practised;" and he adds, that we shall probably have reason to consider "this, compared with the later productions, in the same light as we view the juvenile works of a Milton, or the first manner of a Raphael or other celebrated painter." In a letter to Hutcheson, Hume merely speaks of this article as "somewhat abusive;" so that his vanity, being young and callow, seems to have been correspondingly wide-mouthed and hard to satiate.

[1] Burton, *Life*, vol. i. p. 109.

It must be confessed that, on this occasion, no
less than on that of his other publications, Hume
exhibits no small share of the craving after mere
notoriety and vulgar success, as distinct from the
pardonable, if not honourable, ambition for solid
and enduring fame, which would have harmonised
better with his philosophy. Indeed, it appears to
be by no means improbable that this peculiarity
of Hume's moral constitution was the cause of his
gradually forsaking philosophical studies, after the
publication of the third part ("On Morals") of the
"Treatise," in 1740, and turning to those political
and historical topics which were likely to yield,
and did in fact yield, a much better return of that
sort of success which his soul loved. The
"Philosophical Essays Concerning the Human
Understanding," which afterwards became the
"Inquiry," is not much more than an abridgment
and recast, for popular use, of parts of the
"Treatise," with the addition of the essays on
"Miracles" and on "Necessity." In style, it exhibits
a great improvement on the "Treatise"; but the
substance, if not deteriorated, is certainly not
improved. Hume does not really bring his mature powers to bear upon his early speculations,
in the later work. The crude fruits have not
been ripened, but they have been ruthlessly
pruned away, along with the branches which bore
them. The result is a pretty shrub enough; but
not the tree of knowledge, with its roots firmly

fixed in fact, its branches perennially budding
forth into new truths, which Hume might have
reared. Perhaps, after all, worthy Mrs. Hume
was, in the highest sense, right. Davie was
" wake-minded," not to see that the world of
philosophy was his to overrun and subdue, if he
would but persevere in the work he had begun.
But no—he must needs turn aside for " success " :
and verily he had his reward ; but not the crown
he might have won.

In 1740, Hume seems to have made an
acquaintance which rapidly ripened into a life-long
friendship. Adam Smith was, at that time, a boy
student of seventeen at the University of Glasgow ;
and Hume sends a copy of the " Treatise " to
" Mr. Smith," apparently on the recommendation
of the well-known Hutcheson, Professor of Moral
Philosophy in the university. It is a remarkable
evidence of Adam Smith's early intellectual
development, that a youth of his age should be
thought worthy of such a present.

In 1741 Hume published anonymously, at
Edinburgh, the first volume of " Essays Moral and
Political," which was followed in 1742 by the
second volume.

These pieces are written in an admirable style,
and, though arranged without apparent method, a
system of political philosophy may be gathered
from their contents. Thus the third essay, " That
Politics may be reduced to a Science," defends

that thesis, and dwells on the importance of forms of government.

"So great is the force of laws and of particular forms of government, and so little dependence have they on the humours and tempers of men, that consequences almost as general and certain may sometimes be deduced from them as any which the mathematical sciences afford us."—(III. 15.) (*See* p. 45.)

Hume proceeds to exemplify the evils which inevitably flow from universal suffrage, from aristocratic privilege, and from elective monarchy, by historical examples, and concludes:—

"That an hereditary prince, a nobility without vassals, and a people voting by their representatives, form the best monarchy, aristocracy, and democracy."—(III. 18.)

If we reflect that the following passage of the same essay was written nearly a century and a half ago, it would seem that whatever other changes may have taken place, political warfare remains *in statu quo* :—

"Those who either attack or defend a minister in such a government as ours, where the utmost liberty is allowed, always carry matters to an extreme, and exaggerate his merit or demerit with regard to the public. His enemies are sure to charge him with the greatest enormities, both in domestic and foreign management; and there is no meanness or crime, of which, in their judgment, he is not capable. Unnecessary wars, scandalous treaties, profusion of public treasure, oppressive taxes, every kind of maladministration is ascribed to him. To aggravate the charge, his pernicious conduct, it is said, will extend its baneful influence even to posterity, by undermining

the best constitution in the world, and disordering that wise
system of laws, institutions, and customs, by which our
ancestors, during so many centuries, have been so happily
governed. He is not only a wicked minister in himself, but
has removed every security provided against wicked ministers
for the future.

"On the other hand, the partisans of the minister make his
panegyric rise as high as the accusation against him, and
celebrate his wise, steady, and moderate conduct in every part
of his administration. The honour and interest of the nation
supported abroad, public credit maintained at home, persecution
restrained, faction subdued : the merit of all these blessings is
ascribed solely to the minister. At the same time, he crowns
all his other merits by a religious care of the best government
in the world, which he has preserved in all its parts, and has
transmitted entire, to be the happiness and security of the
latest posterity."—(III. 26.)

Hume sagely remarks that the panegyric and
the accusation cannot both be true ; and, that what
truth there may be in either, rather tends to show
that our much-vaunted constitution does not fulfil
its chief object, which is to provide a remedy
against maladministration. And if it does not—

"we are rather beholden to any minister who undermines it
and affords us the opportunity of erecting a better in its
place."—(III. 28.)

The fifth Essay discusses the " Origin of
Government " :—

"Man, born in a family, is compelled to maintain society
from necessity, from natural inclination, and from habit. The
same creature, in his farther progress, is engaged to establish
political society, in order to administer justice, without which
there can be no peace among them, nor safety, nor mutual

intercourse. We are therefore to look upon all the vast
apparatus of our government, as having ultimately no other
object or purpose but the distribution of justice, or, in other
words, the support of the twelve judges. Kings and parliaments, fleets and armies, officers of the court and revenue, am-
bassadors, ministers and privy councillors, are all subordinate
in the end to this part of administration. Even the clergy, as
their duty leads them to inculcate morality, may justly be
thought, so far as regards this world, to have no other useful
object of their institution."—(III. 37.)

The police theory of government has never been
stated more tersely: and, if there were only one
state in the world ; and if we could be certain by
intuition, or by the aid of revelation, that it is
wrong for society, as a corporate body, to do
anything for the improvement of its members and,
thereby, indirectly support the twelve judges, no
objection could be raised to it.

Unfortunately the existence of rival or inimical
nations furnishes "kings and parliaments, fleets
and armies," with a good deal of occupation
beyond the support of the twelve judges; and,
though the proposition that the State has no
business to meddle with anything but the administration of justice, seems sometimes to be
regarded as an axiom, it can hardly be said to
be intuitively certain, inasmuch as a great many
people absolutely repudiate it ; while, as yet, the
attempt to give it the authority of a revelation
has not been made.

As Hume says with profound truth in the

fourth Essay, "On the First Principles of Government":

"As force is always on the side of the governed, the governors have nothing to support them but opinion. It is, therefore, on opinion only that government is founded; and this maxim extends to the most despotic and most military governments, as well as to the most free and the most popular."—(III. 31.)

But if the whole fabric of social organisation rests on opinion, it may surely be fairly argued that, in the interests of self-preservation, if for no better reason, society has a right to see that the means of forming just opinions are placed within the reach of every one of its members; and, therefore, that due provision for education, at any rate, is a right and, indeed, a duty, of the state.

The three opinions upon which all government, or the authority of the few over the many, is founded, says Hume, are public interest, right to power, and right to property. No government can permanently exist, unless the majority of the citizens, who are the ultimate depositary of Force, are convinced that it serves the general interest, that it has lawful authority, and that it respects individual rights:—

"A government may endure for several ages, though the balance of power and the balance of property do not coincide . . . . But where the original constitution allows any share of power, though small, to an order of men who possess a large share of property, it is easy for them gradually to stretch their authority, and bring the balance of power to coincide with that of property. This has been the case with the House of Commons in England."—(III. 34.)

Hume then points out that, in his time, the authority of the Commons was by no means equivalent to the property and power it represented, and proceeds :—

"Were the members obliged to receive instructions from their constituents, like the Dutch deputies, this would entirely alter the case ; and if such immense power and riches as those of all the Commons of Great Britain, were brought into the scale, it is not easy to conceive that the crown could either influence that multitude of people, or withstand that balance of property. It is true, the crown has great influence over the collective body in the elections of members ; but were this influence, which at present is only exerted once in seven years, to be employed in bringing over the people to every vote, it would soon be wasted, and no skill, popularity, or revenue could support it. I must, therefore, be of opinion that an alteration in this particular would introduce a total alteration in our government, would soon reduce it to a pure republic ; and, perhaps, to a republic of no inconvenient form."—(III. 35.)

Viewed by the light of subsequent events, this is surely a very remarkable example of political sagacity. The members of the House of Commons are not yet delegates; but, with the widening of, the suffrage and the rapidly increasing tendency to drill and organise the electorate, and to exact definite pledges from candidates, they are rapidly becoming, if not delegates, at least attorneys for committees of electors. The same causes are constantly tending to exclude men, who combine a keen sense of self-respect with large intellectual capacity, from a position in which the one is as constantly offended, as the other is neutralised.

Notwithstanding the attempt of George the Third to resuscitate the royal authority, Hume's foresight has been so completely justified that no one now dreams of the crown exerting the slightest influence upon elections.

In the seventh Essay, Hume raises a very interesting discussion as to the probable ultimate result of the forces which were at work in the British Constitution in the first part of the eighteenth century :—

"There has been a sudden and sensible change in the opinions of men, within these last fifty years, by the progress of learning and of liberty. Most people in this island have divested themselves of all superstitious reverence to names and authority ; the clergy have much lost their credit; their pretensions and doctrines have been much ridiculed ; and even religion can scarcely support itself in the world. The mere name of *king* commands little respect ; and to talk of a king as God's vicegerent on earth, or to give him any of those magnificent titles which formerly dazzled mankind, would but excite laughter in every one."—(III. 54.)

In fact, at the present day, the danger to monarchy in Britain would appear to lie, not in increasing love for equality, for which, except as regards the law, Englishmen have never cared, but rather entertain an aversion; nor in any abstract democratic theories, upon which the mass of Englishmen pour the contempt with which they view theories in general; but in the constantly increasing tendency of monarchy to become slightly absurd, from the ever-widening

discrepancy between modern political ideas and
the theory of kingship. As Hume observes,
even in his time, people had left off making
believe that a king was a different species of man
from other men; and, since his day, more and
more such make-believes have become impossible;
until the maintenance of kingship in coming
generations seems likely to depend, entirely, upon
whether it is the general opinion, that a hereditary
president of our virtual republic will serve the
general interest better than an elective one or
not. The tendency of public feeling in this
direction is patent, but it does not follow that
a republic is to be the final stage of our government. In fact Hume thinks not :—

"It is well known, that every government must come to a
period, and that death is unavoidable to the political, as well
as to the animal body. But, as one kind of death may be
preferable to another, it may be inquired, whether it be more
desirable for the British constitution to terminate in a popular
government, or in an absolute monarchy? Here, I would
frankly declare, that though liberty be preferable to slavery, in
almost every case ; yet I should rather wish to see an absolute
monarch than a republic in this island. For let us consider
what kind of republic we have reason to expect. The question
is not concerning any fine imaginary republic of which a man
forms a plan in his closet. There is no doubt but a popular
government may be imagined more perfect than an absolute
monarchy, or even than our present constitution. But what
reason have we to expect that any such government will ever be
established in Great Britain, upon the dissolution of our
monarchy? If any single person acquire power enough to take
our constitution to pieces, and put it up anew, he is really

an absolute monarch ; and we have already had an instance of
this kind, sufficient to convince us, that such a person will
never resign his power, or establish any free government.
Matters, therefore, must be trusted to their natural progress
and operation ; and the House of Commons, according to its
present constitution, must be the only legislature in such a
popular government. The inconveniences attending such a
situation of affairs present themselves by thousands. If the
House of Commons, in such a case, ever dissolve itself, which
is not to be expected, we may look for a civil war every
election. If it continue itself, we shall suffer all the tyranny
of a faction subdivided into new factions. And, as such a
violent government cannot long subsist, we shall at last, after
many convulsions and civil wars, find repose in absolute
monarchy, which it would have been happier for us to have
established peaceably from the beginning. Absolute monarchy,
therefore, is the easiest death, the true *Euthanasia* of the
British constitution.

" Thus if we have more reason to be jealous of monarchy,
because the danger is more imminent from that quarter ; we
have also reason to be more jealous of popular government, because that danger is more terrible. This may teach us a lesson
of moderation in all our political controversies."—(III. 55).

One may admire the sagacity of these speculations, and the force and clearness with which they
are expressed, without altogether agreeing with
them. That an analogy between the social and
bodily organism exists, and is, in many respects,
clear and full of instructive suggestion, is undeniable. Yet a state answers, not to an individual,
but to a generic type ; and there is no reason, in
the nature of things, why any generic type should
die out. The type of the pearly *Nautilus*, highly
organised as it is, has persisted with but little

change from the Silurian epoch till now ; and, so
long as terrestrial conditions remain approximately similar to what they are at present, there
is no more reason why it should cease to exist in
the next, than in the past, hundred million years
or so. The true ground for doubting the possibility of the establishment of absolute monarchy
in Britain is, that opinion seems to have passed
through, and left far behind, the stage at which
such a change would be possible ; and the true
reason for doubting the permanency of a republic,
if it is ever established, lies in the fact, that a
republic requires for its maintenance a far higher
standard of morality and of intelligence in the
members of the state than any other form of
government. Samuel gave the Israelites a king
because they were not righteous enough to do
without one, with a pretty plain warning of what
they were to expect from the gift. And, up to
this time, the progress of such republics as have
been established in the world has not been such,
as to lead to any confident expectation that their
foundation is laid on a sufficiently secure subsoil
of public spirit, morality, and intelligence. On
the contrary, they exhibit examples of personal
corruption and of political profligacy as fine as any
hotbed of despotism has ever produced ; while
they fail in the primary duty of the administration of justice, as none but an effete despotism
has ever failed.

Hume has been accused of departing, in his old age, from the liberal principles of his youth ; and, no doubt, he was careful, in the later editions of the " Essays," to expunge everything that savoured of democratic tendencies. But the passage just quoted shows that this was no recantation, but simply a confirmation, by his experience of one of the most debased periods of English history, of those evil tendencies attendant on popular government, of which, from the first, he was fully aware.

In the ninth essay, "On the Parties of Great Britain," there occurs a passage which, while it affords evidence of the marvellous change which has taken place in the social condition of Scotland since 1741, contains an assertion respecting the state of the Jacobite party at that time, which at first seems surprising :—

" As violent things have not commonly so long a duration as moderate, we actually find that the Jacobite party is almost entirely vanished from among us, and that the distinction of *Court* and *Country*, which is but creeping in at London, is the only one that is ever mentioned in this kingdom. Beside the violence and openness of the Jacobite party, another reason has perhaps contributed to produce so sudden and so visible an alteration in this part of Britain. There are only two ranks of men among us ; gentlemen who have some fortune and education, and the meanest slaving poor ; without any considerable number of that middling rank of men, which abound more in England, both in cities and in the country, than in any other part of the world. The slaving poor are incapable of any principles ; gentlemen may be converted to true principles, by time and experience. The middling rank of men have curiosity and knowledge enough to form principles, but not

enough to form true ones, or correct any prejudices that they
may have imbibed. And it is among the middling rank of
people that Tory principles do at present prevail most in
England."—(III. 80, *note.*)

Considering that the Jacobite rebellion of 1745
broke out only four years after this essay was
published, the assertion that the Jacobite party
had "almost entirely vanished in 1741" sounds
strange enough : and the passage which contains
it is omitted in the third edition of the "Essays,"
published in 1748. Nevertheless, Hume was
probably right, as the outbreak of '45 was little
better than a Highland raid, and the Pretender
obtained no important following in the Lowlands.

No less curious, in comparison with what would
be said nowadays, is Hume's remark in the essay
on the " Rise of the Arts and Sciences " that—

"The English are become sensible of the scandalous licentiousness of their stage from the example of the French decency
and morals."—(III. 135.)

And it is perhaps as surprising to be told, by a
man of Hume's literary power, that the first polite
prose in the English language was written by
Swift. Locke and Temple (with whom Sprat is
astoundingly conjoined) " knew too little of the
rules of art to be esteemed elegant writers," and
the prose of Bacon, Harrington, and Milton is
" altogether stiff and pedantic." Hobbes, who
whether he should be called a " polite " writer or

not, is a master of vigorous English; Clarendon, Addison, and Steele (the last two, surely, were "polite" writers in all conscience) are not mentioned.

On the subject of "National Character," about which more nonsense, and often very mischievous nonsense, has been and is talked than upon any other topic, Hume's observations are full of sense and shrewdness. He distinguishes between the *moral* and the *physical* causes of national character, enumerating under the former—

"The nature of the government, the revolutions of public affairs, the plenty or penury in which people live, the situation of the nation with regard to its neighbours, and such like circumstances."—(III. 225.)

and under the latter :—

"Those qualities of the air and climate, which are supposed to work insensibly on the temper, by altering the tone and habit of the body, and giving a particular complexion, which, though reflexion and reason may sometimes overcome it, will yet prevail among the generality of mankind, and have an influence on their manners."—(III. 225.)

While admitting and exemplifying the great influence of moral causes, Hume remarks—

"As to physical causes, I am inclined to doubt altogether of their operation in this particular; nor do I think that men owe anything of their temper or genius to the air, food, or climate."—(III. 227.)

Hume certainly would not have accepted the "rice theory" in explanation of the social state of

the Hindoos; and, it may be safely assumed, that
he would not have had recourse to the circum-
ambience of the " melancholy main " to account
for the troublous history of Ireland. He supports
his views by a variety of strong arguments,
among which, at the present conjuncture, it is
worth noting that the following occurs—

"Where any accident, as a difference in language or religion,
keeps two nations, inhabiting the same country, from mixing
with one another, they will preserve during several centuries
a distinct and even opposite set of manners. The integrity,
gravity, and bravery of the Turks, form an exact contrast to
the deceit, levity, and cowardice of the modern Greeks."—
(III. 233.)

The question of the influence of race, which
plays so great a part in modern political speculations, was hardly broached in Hume's time, but he
had an inkling of its importance :—

"I am apt to suspect the Negroes to be naturally inferior
to the Whites. There scarcely ever was a civilised nation of
that complexion, nor even any individual, eminent either in
action or speculation. . . . Such a uniform and constant
difference [between the negroes and the whites] could not
happen in so many countries and ages, if nature had not made
an original distinction between these breeds of men. . . .
In Jamaica, indeed, they talk of one Negro as a man of
parts and learning ; but it is likely he is admired for slender
accomplishments, like a parrot who speaks a few words
plainly."—(III. 236.)

The "Essays" met with the success they deserved.
Hume wrote to Henry Home in June, 1742 :—

"The Essays are all sold in London, as I am informed by two letters from English gentlemen of my acquaintance. There is a demand for them ; and, as one of them tells me, Innys, the great bookseller in Paul's Churchyard, wonders there is not a new edition, for he cannot find copies for his customers. I am also told that Dr. Butler has everywhere recommended them ; so that I hope that they will have some success."

Hume had sent Butler a copy of the " Treatise " and had called upon him, in London, but he was out of town; and being shortly afterwards made Bishop of Bristol, Hume seems to have thought that further advances on his part might not be well received.

Greatly comforted by this measure of success, Hume remained at Ninewells, rubbing up his Greek, until 1745; when, at the mature age of thirty-four, he made his entry into practical life, by becoming bear-leader to the Marquis of Annandale, a young nobleman of feeble body and feebler mind. As might have been predicted, this venture was not more fortunate than his previous ones; and, after a year's endurance, diversified latterly with pecuniary squabbles, in which Hume's tenacity about a somewhat small claim is remarkable, the engagement came to an end.

CHAPTER II

LATER YEARS: THE HISTORY OF ENGLAND

In 1744, Hume's friends had endeavoured to procure his nomination to the Chair of "Ethics and pneumatic philosophy"[1] in the University of Edinburgh. About this matter he writes to his friend William Mure:—

> "The accusation of heresy, deism, scepticism, atheism, &c., &c., &c. was started against me; but never took, being bore down by the contrary authority of all the good company in town."

If the "good company in town" bore down the first three of these charges, it is to be hoped, for the sake of their veracity, that they knew their candidate chiefly as the very good company that he always was; and had paid as little attention, as good company usually does, to so solid a work as the "Treatise." Hume expresses a naïve

[1] "Pneumatic philosophy" must not be confounded with the theory of elastic fluids; though, as Scottish chairs have, before now, combined natural with civil history, the mistake would be pardonable.

surprise, not unmixed with indignation, that
Hutcheson and Leechman, both clergymen and
sincere, though liberal, professors of orthodoxy,
should have expressed doubts as to his fitness for
becoming a professedly presbyterian teacher of
presbyterian youth. The town council, however,
would not have him, and filled up the place with
a safe nobody.

In May, 1746, a new prospect opened. General
St. Clair was appointed to the command of an
expedition to Canada, and he invited Hume, at a
week's notice, to be his secretary; to which office
that of judge advocate was afterwards added.

Hume writes to a friend : " The office is very
genteel, 10s. a day, perquisites, and no expenses ; "
and, to another, he speculates on the chance of
procuring a company in an American regiment.
" But this I build not on, nor indeed am I very
fond of it," he adds ; and this was fortunate, for
the expedition, after dawdling away the summer
in port, was suddenly diverted to an attack on
L'Orient, where it achieved a huge failure and
returned ignominiously to England.

A letter to Henry Home, written when this un-
lucky expedition was recalled, shows that Hume
had already seriously turned his attention to history. Referring to an invitation to go over to
Flanders with the General, he says :

" Had I any fortune which would give me a prospect of
leisure and opportunity to prosecute my *historical projects*,

nothing could be more useful to me, and I should pick up more literary knowledge in one campaign by being in the General's family, and being introduced frequently to the Duke's, than most officers could do after many years' service. But to what can all this serve ? I am a philosopher, and so I suppose must continue."

But this vaticination was shortly to prove erroneous. Hume seems to have made a very favourable impression on General St. Clair, as he did upon every one with whom he came into personal contact; for, being charged with a mission to the Court of Turin, in 1748, the General insisted upon the appointment of Hume as his secretary. He further made him one of his aides-de-camp ; so that the philosopher was obliged to encase his more than portly, and by no means elegant, figure in a military uniform. Lord Charlemont, who met him at Turin, says he was " disguised in scarlet," and that he wore his uniform "like a grocer of the train-bands." Hume, always ready for a joke at his own expense, tells of the considerate kindness with which, at a reception at Vienna, the Empress-dowager released him and his friends from the necessity of walking backwards. " We esteemed ourselves very much obliged to her for this attention, especially my companions, who were desperately afraid of my falling on them and crushing them."

Notwithstanding the many attractions of this appointment, Hume writes that he leaves home "with infinite regret, where I had treasured up stores of study and plans of thinking for many years;" and his only consolation is that the opportunity of becoming conversant with state affairs may be profitable :—

"I shall have an opportunity of seeing courts and camps; and if I can afterward be so happy as to attain leisure and other opportunities, this knowledge may even turn to account to me as a man of letters, which I confess has always been the sole object of my ambition. I have long had an intention, in my riper years, of composing some history; and I question not but some greater experience in the operations of the field and the intrigues of the cabinet will be requisite, in order to enable me to speak with judgment on these subjects."

Hume returned to London in 1749, and during his stay there, his mother died, to his heartfelt sorrow. A curious story in connection with this event is told by Dr. Carlyle, who knew Hume well, and whose authority is perfectly trustworthy.

"Mr. Boyle hearing of it, soon after went to his apartment, for they lodged in the same house, where he found him in the deepest affliction and in a flood of tears. After the usual topics and condolences Mr. Boyle said to him, 'My friend, you owe this uncommon grief to having thrown off the principles of religion : for if you had not, you would have been consoled with the firm belief that the good lady, who was not only the best of mothers, but the most pious of Christians, was completely happy in the realms of the just.' To which David replied, 'Though I throw out my speculations to entertain the learned and metaphysical world, yet in other things I do not think so differently from the rest of the world as you imagine.'"

If Hume had told this story to Dr. Carlyle, the latter would have said so; it must therefore have come from Mr. Boyle; and one would like to have the opportunity of cross-examining that gentleman as to Hume's exact words and their context, before implicitly accepting his version of the conversation. Mr. Boyle's experience of mankind must have been small, if he had not seen the firmest of believers overwhelmed with grief by a like loss, and as completely inconsolable. Hume may have thrown off Mr. Boyle's " principles of religion," but he was none the less a very honest man, perfectly open and candid, and the last person to use ambiguous phraseology among his friends; unless, indeed, he saw no other way of putting a stop to the intrusion of unmannerly twaddle amongst the bitter-sweet memories stirred in his affectionate nature by so heavy a blow.

The " Philosophical Essays " or " Inquiry " was published in 1748, while Hume was away with General St. Clair, and, on his return to England, he had the mortification to find it overlooked in the hubbub caused by Middleton's " Free Inquiry," and its bold handling of the topic of the " Essay on Miracles," by which Hume doubtless expected the public to be startled.

Between 1749 and 1751, Hume resided at Ninewells, with his brother and sister, and busied himself with the composition of his most finished, if not his most important works, the " Dialogues

on Natural Religion," the "Inquiry Concerning
the Principles of Morals," and the "Political
Discourses."

"The Dialogues on Natural Religion" were
touched and re-touched, at intervals, for a quarter
of a century, and were not published till after
Hume's death : but the " Inquiry Concerning the
Principles of Morals" appeared in 1751, and the
"Political Discourses " in 1752. Full reference
will be made to the two former in the exposition
of Hume's philosophical views. The last has been
well said to be the "cradle of political economy :
and much as that science has been investigated
and expounded in later times, these earliest,
shortest, and simplest developments of its principles are still read with delight even by those
who are masters of all the literature of this great
subject."[1]

The " Wealth of Nations," the masterpiece of
Hume's close friend, Adam Smith, it must be
remembered, did not appear before 1776, so that,
in political economy, no less than in philosophy,
Hume was an original, a daring, and a fertile
innovator.

The "Political Essays" had a great and rapid
success ; translated into French in 1753, and
again in 1754, they conferred a European reputation upon their author; and, what was more to

[1] Burton's *Life of David Hume*, i. p. 354.

the purpose, influenced the later French school of
economists of the eighteenth century.

By this time, Hume had not only attained a
high reputation in the world of letters, but he
considered himself a man of independent fortune.
His frugal habits had enabled him to accumulate
£1,000, and he tells Michael Ramsay in 1751 :—

"While interest remains as at present, I have £50 a year, a
hundred pounds worth of books, great store of linens and fine
clothes, and near £100 in my pocket; along with order,
frugality, a strong spirit of independency, good health, a
contented humour, and an unabated love of study. In these
circumstances I must esteem myself one of the happy and
fortunate ; and so far from being willing to draw my ticket
over again in the lottery of life, there are very few prizes with
which I would make an exchange. After some deliberation,
I am resolved to settle in Edinburgh, and hope I shall be able
with these revenues to say with Horace :—

> 'Est bona librorum et provisæ frugis in annum
> Copia.'"

It would be difficult to find a better example of
the honourable independence and cheerful self-
reliance which should distinguish a man of letters,
and which characterised Hume throughout his
career. By honourable effort, the boy's noble
ideal of life, became the man's reality; and, at
forty, Hume had the happiness of finding that he
had not wasted his youth in the pursuit of
illusions, but that "the solid certainty of waking
bliss" lay before him in the free play of his powers
in their appropriate sphere.

In 1751, Hume removed to Edinburgh and took up his abode on a flat in one of those prodigious houses in the Lawnmarket, which still excite the admiration of tourists; afterwards moving to a house in the Canongate. His sister joined him, adding £30 a year to the common stock; and, in one of his charmingly playful letters to Dr. Clephane, he thus describes his establishment, in 1753

"I shall exult and triumph to you a little that I have now at last—being turned of forty, to my own honour, to that of learning, and to that of the present age—arrived at the dignity of being a householder.

"About seven months ago, I got a house of my own, and completed a regular family, consisting of a head, viz., myself, and two inferior members, a maid and a cat. My sister has since joined me, and keeps me company. With frugality, I can reach, I find, cleanliness, warmth, light, plenty, and contentment. What would you have more? Independence? I have it in a supreme degree. Honour? That is not altogether wanting. Grace? That will come in time. A wife? That is none of the indispensable requisites of life. Books? That *is* one of them; and I have more than I can use. In short, I cannot find any pleasure of consequence which I am not possessed of in a greater or less degree: and, without any great effort of philosophy, I may be easy and satisfied.

"As there is no happiness without occupation, I have begun a work which will occupy me several years, and which yields me much satisfaction. 'Tis a History of Britain from the Union of the Crowns to the present time. I have already finished the reign of King James. My friends flatter me (by this I mean that they don't flatter me) that I have succeeded."

In 1752, the Faculty of Advocates elected Hume their librarian, an office which, though it

yielded little emolument—the salary was only
forty pounds a year—was valuable as it placed
the resources of a large library at his disposal.
The proposal to give Hume even this paltry place
caused a great outcry, on the old score of infidel-
ity. But as Hume writes, in a jubilant letter to
Clephane (February 4, 1752):—

"I carried the election by a considerable majority. . . .
What is more extraordinary, the cry of religion could not
hinder the ladies from being violently my partisans, and I owe
my success in a great measure to their solicitations. One has
broke off all commerce with her lover because he voted against
me! And Mr. Lockhart, in a speech to the Faculty, said there
was no walking the streets, nor even enjoying one's own fire-
side, on account of their importunate zeal. The town says that
even his bed was not safe for him, though his wife was cousin-
german to my antagonist.

"'Twas vulgarly given out that the contest was between
Deists and Christians, and when the news of my success came
to the playhouse, the whisper rose that the Christians were
defeated. Are you not surprised that we could keep our popularity, notwithstanding this imputation, which my friends could
not deny to be well founded?"

It would seem that the "good company" was
less enterprising in its asseverations in this canvass
than in the last.

The first volume of the "History of Great
Britain, containing the reign of James I. and
Charles I.," was published in 1754. At first, the
sale was large, especially in Edinburgh, and if
notoriety *per se* was Hume's object, he attained it.

But he liked applause as well as fame, and, to his bitter disappointment, he says :—

> "I was assailed by one cry of reproach, disapprobation, and even detestation : English, Scotch, and Irish, Whig and Tory, Churchman and Sectary, Freethinker and Religionist, Patriot and Courtier, united in their rage against the man who had presumed to shed a generous tear for the fate of Charles I. and the Earl of Strafford ; and after the first ebullitions of their fury were over, what was still more mortifying, the book seemed to fall into oblivion. Mr. Millar told me that in a twelvemonth he sold only forty-five copies of it. I scarcely, indeed, heard of one man in the three kingdoms, considerable for rank or letters, that could endure the book. I must only except the primate of England, Dr. Herring, and the primate of Ireland, Dr. Stone, which seem two odd exceptions. These dignified prelates separately sent me messages not to be discouraged."

It certainly is odd to think of David Hume being comforted in his affliction by the independent and spontaneous sympathy of a pair of archbishops. But the instincts of the dignified prelates guided them rightly; for, as the great painter of English history in Whig pigments has been careful to point out,[1] Hume's historical picture, though a great work, drawn by a master hand, has all the lights Tory, and all the shades Whig.

Hume's ecclesiastical enemies seem to have thought that their opportunity had now arrived ; and an attempt was made to get the General

[1] Lord Macaulay, Article on History, *Edinburgh Review*, vol. lxvii.

Assembly of 1756 to appoint a committee to inquire into his writings. But, after a keen debate, the proposal was rejected by fifty votes to seventeen. Hume does not appear to have troubled himself about the matter, and does not even think it worth mention in " My Own Life."

In 1756 he tells Clephane that he is worth £1,600 sterling, and consequently master of an income which must have been wealth to a man of his frugal habits. In the same year, he published the second volume of the " History," which met with a much better reception than the first; and, in 1757, one of his most remarkable works, the " Natural History of Religion," appeared. In the same year, he resigned his office of librarian to the Faculty of Advocates, and he projected removal to London, probably to superintend the publication of the additional volume of the " History."

" I shall certainly be in London next summer ; and probably to remain there during life : at least, if I can settle myself to my mind, which I beg you to have an eye to. A room in a sober discreet family, who would not be averse to admit a sober, discreet, virtuous, regular, quiet, goodnatured man of a bad character—such a room, I say, would suit me extremely." [1]

The promised visit took place in the latter part of the year 1758, and he remained in the

[1] Letter to Clephane, 3rd September, 1757.

metropolis for the greater part of 1759. The two volumes of the "History of England under the House of Tudor" were published in London, shortly after Hume's return to Edinburgh; and, according to his own account, they raised almost as great a clamour as the first two had done.

Busily occupied with the continuation of his historical labours, Hume remained in Edinburgh until 1763; when, at the request of Lord Hertford, who was going as ambassador to France, he was appointed to the embassy; with the promise of the secretaryship, and, in the meanwhile, performing the duties of that office. At first, Hume declined the offer; but, as it was particularly honourable to so well abused a man, on account of Lord Hertford's high reputation for virtue and piety,[1] and no less advantageous by reason of the increase of fortune which it secured to him, he eventually accepted it.

In France, Hume's reputation stood far higher than in Britain; several of his works had been translated; he had exchanged letters with Montesquieu and with Helvetius; Rousseau had appealed to him; and the charming Madame de Boufflers had drawn him into a correspondence,

---

[1] "You must know that Lord Hertford has so high a character for piety, that his taking me by the hand is a kind of regeneration to me, and all past offences are now wiped off. But all these views are trifling to one of my age and temper."— *Hume to Edmonstone*, 9th January, 1764. Lord Hertford had procured him a pension of £200 a year for life from the King, and the secretaryship was worth £1,000 a year.

marked by almost passionate enthusiasm on her part, and as fair an imitation of enthusiasm as Hume was capable of, on his. In the extraordinary mixture of learning, wit, humanity, frivolity, and profligacy which then characterised the highest French society, a new sensation was worth anything, and it mattered little whether the cause thereof was a philosopher or a poodle; so Hume had a great success in the Parisian world. Great nobles fêted him, and great ladies were not content unless the "gros David" was to be seen at their receptions, and in their boxes at the theatre. "At the opera his broad unmeaning face was usually to be seen *entre deux jolis minois*," says Lord Charlemont.[1] Hume's cool head was by no means turned; but he took the goods the gods provided with much satisfaction; and everywhere won golden opinions by his unaffected good sense and thorough kindness of heart.

Over all this part of Hume's career, as over the surprising episode of the quarrel with Rousseau, if that can be called quarrel which was lunatic

---

[1] Madame d'Epinay gives a ludicrous account of Hume's performance when pressed into a *tableau*, as a Sultan between two slaves, personated for the occasion by two of the prettiest women in Paris :—

"Il les regarde attentivement, *il se frappe le ventre* et les genoux à plusieurs reprises et ne trouve jamais autre chose à leur dire que. *Eh bien! mes demoiselles.—Eh bien! vous voilà donc. . . . Eh bien! vous voilà . . . vous voilà ici?* Cette phrase dura un quart d'heure sans qu'il pût en sortir. Une d'elles se leva d'impatience : Ah, dit-elle, je m'en étois bien doutée, cet homme n'est bon qu'à manger du veau !"—Burton's *Life of Hume*, vol. ii. p. 224.

malignity on Rousseau's side and thorough
generosity and patience on Hume's, I may pass
lightly. The story is admirably told by Mr.
Burton, to whose volumes I refer the reader.
Nor need I dwell upon Hume's short tenure of
office in London, as Under-Secretary of State,
between 1767 and 1769. Success and wealth are
rarely interesting, and Hume's case is no exception
to the rule.

According to his own description the cares of
official life were not overwhelming.

"My way of life here is very uniform and by no means
disagreeable. I have all the forenoon in the Secretary's house,
from ten till three, when there arrive from time to time
messengers that bring me all the secrets of the kingdom, and,
indeed, of Europe, Asia, Africa, and America. I am seldom
hurried ; but have leisure at intervals to take up a book, or
write a private letter, or converse with a friend that may call
for me ; and from dinner to bed-time is all my own. If you
add to this that the person with whom I have the chief, if not
only, transactions, is the most reasonable, equal-tempered, and
gentleman-like man imaginable, and Lady Aylesbury the same,
you will certainly think I have no reason to complain ; and I
am far from complaining. I only shall not regret when my
duty is over ; because to me the situation can lead to nothing,
at least in all probability ; and reading, and sauntering, and
lounging, and dozing, which I call thinking, is my supreme
happiness—I mean my full contentment."

Hume's duty was soon over, and he returned to
Edinburgh in 1769, "very opulent" in the
possession of £1,000 a year, and determined to
take what remained to him of life pleasantly

and easily. In October, 1769, he writes to Elliot:—

"I have been settled here two months, and am here body and soul, without casting the least thought of regret to London, or even to Paris . . . I live still, and must for a twelvemonth, in my old house in James's Court, which is very cheerful and even elegant, but too small to display my great talent for cookery, the science to which I intend to addict the remaining years of my life. I have just now lying on the table before me a receipt for making *soupe à la reine*, copied with my own hand ; for beef and cabbage (a charming dish) and old mutton and old claret nobody excels me. I make also sheep's-head broth in a manner that Mr. Keith speaks of for eight days after ; and the Duc de Nivernois would bind himself apprentice to my lass to learn it. I have already sent a challenge to David Moncrieff : you will see that in a twelvemonth he will take to the writing of history, the field I have deserted ; for as to the giving of dinners, he can now have no further pretensions. I should have made a very bad use of my abode in Paris if I could not get the better of a mere provincial like him. All my friends encourage me in this ambition ; as thinking it will redound very much to my honour."

In 1770, Hume built himself a house in the new town of Edinburgh, which was then springing up. It was the first house in the street, and a frolicsome young lady chalked upon the wall "St. David's Street." Hume's servant complained to her master, who replied, "Never mind, lassie, many a better man has been made a saint of before," and the street retains its title to this day.

In the following six years, the house in St. David's Street was the centre of the accomplished

and refined society which then distinguished
Edinburgh. Adam Smith, Blair, and Ferguson
were within easy reach; and what remains of
Hume's correspondence with Sir Gilbert Elliot,
Colonel Edmonstone, and Mrs. Cockburn gives
pleasant glimpses of his social surroundings, and
enables us to understand his contentment with
his absence from the more perturbed, if more
brilliant, worlds of Paris and London.

Towards London, Londoners, and indeed
Englishmen in general, Hume entertained a
dislike, mingled with contempt, which was as
nearly rancorous as any emotion of his could be.
During his residence in Paris, in 1764 and 1765,
he writes to Blair :—

"The taste for literature is neither decayed nor depraved
here, as with the barbarians who inhabit the banks of the
Thames."

And he speaks of the "general regard paid to
genius and learning" in France as one of the
points in which it most differs from England.
Ten years later, he cannot even thank Gibbon for
his History without the lefthanded compliment,
that he should never have expected such an
excellent work from the pen of an Englishman.
Early in 1765, Hume writes to Millar :—

"The rage and prejudice of parties frighten me, and above
all, this rage against the Scots, which is so dishonourable, and
indeed so infamous, to the English nation. We hear that it

increases every day without the least appearance of provocation
on our part. It has frequently made me resolve never in my
life to set foot on English ground. I dread, if I should undertake a more modern history, the impertinence and ill manners
to which it would expose me ; and I was willing to know from
you whether former prejudices had so far subsided as to ensure
me of a good reception."

His fears were kindly appeased by Millar's
assurance that the English were not prejudiced
against the Scots in general, but against the
particular Scot, Lord Bute, who was supposed to
be the guide, philosopher, and friend, of both the
King and his mother.

To care nothing about literature, to dislike
Scotchmen, and to be insensible to the merits of
David Hume, was a combination of iniquities on
the part of the English nation, which would have
been amply sufficient to ruffle the temper of the
philosophic historian, who, without being foolishly
vain, had certainly no need of what has been said
to be the one form of prayer in which his country-
men, torn as they are by theological differences,
agree ; " Lord ! gie us a gude conceit o' oursels."
But when, to all this, these same Southrons
added a passionate admiration for Lord Chatham,
who was in Hume's eyes a charlatan ; and filled
up the cup of their abominations by cheering for
" Wilkes and Liberty," Hume's wrath knew no
bounds, and, between 1768 and 1770, he pours a
perfect Jeremiad into the bosom of his friend Sir
Gilbert Elliot.

"Oh! how I long to see America and the East Indies revolted, totally and finally—the revenue reduced to half—public credit fully discredited by bankruptcy—the third of London in ruins, and the rascally mob subdued! I think I am not too old to despair of being witness to all these blessings.

"I am delighted to see the daily and hourly progress of madness and folly and wickedness in England. The consummation of these qualities are the true ingredients for making a fine narrative in history, especially if followed by some signal and ruinous convulsion—as I hope will soon be the case with that pernicious people!"

Even from the secure haven of James's Court, the maledictions continue to pour forth :—

"Nothing but a rebellion and bloodshed will open the eyes of that deluded people; though were they alone concerned, I think it is no matter what becomes of them. . . . Our government has become a chimera, and is too perfect, in point of liberty, for so rude a beast as an Englishman; who is a man, a bad animal too, corrupted by above a century of licentiousness. The misfortune is that this liberty can scarcely be retrenched without danger of being entirely lost; at least the fatal effects of licentiousness must first be made palpable by some extreme mischief resulting from it. I may wish that the catastrophe should rather fall on our posterity, but it hastens on with such large strides as to leave little room for hope.

"I am running over again the last edition of my History, in order to correct it still further. I either soften or expunge many villainous seditious Whig strokes which had crept into it. I wish that my indignation at the present madness, encouraged by lies, calumnies, imposture, and every infamous act usual among popular leaders, may not throw me into the opposite extreme."

A wise wish, indeed. Posterity respectfully concurs therein ; and subjects Hume's estimate of England and things English to such modifications as it would probably have undergone had the wish been fulfilled.

In 1775, Hume's health began to fail; and in the spring of the following year, his disorder, which appears to have been hæmorrhage of the bowels, attained such a height that he knew it must be fatal. So he made his will, and wrote "My Own Life," the conclusion of which is one of the most cheerful, simple, and dignified leave-takings of life and all its concerns, extant.

"I now reckon upon a speedy dissolution. I have suffered very little pain from my disorder ; and what is more strange, have, notwithstanding the great decline of my person, never suffered a moment's abatement of spirits ; insomuch that were I to name the period of my life which I should most choose to pass over again, I might be tempted to point to this later period. I possess the same ardour as ever in study and the same gaiety in company ; I consider, besides, that a man of sixty-five, by dying, cuts off only a few years of infirmities ; and though I see many symptoms of my literary reputation's breaking out at last with additional lustre, I know that I could have but few years to enjoy it. It is difficult to be more detached from life than I am at present.

"To conclude historically with my own character, I am, or rather was (for that is the style I must now use in speaking of myself, which emboldens me the more to speak my senti- ments) ; I was, I say, a man of mild dispositions, of command of temper, of an open, social, and cheerful humour, capable of attachment, but little susceptible of enmity, and of great moderation in all my passions. Even my love of literary

fame, my ruling passion, never soured my temper, notwithstanding my frequent disappointments. My company was
not unacceptable to the young and careless, as well as to the
studious and literary ; and as I took a particular pleasure in the
company of modest women, I had no reason to be displeased
with the reception I met with from them. In a word, though
most men any wise eminent, have found reason to complain
of calumny, I never was touched or even attacked by her
baleful tooth ; and though I wantonly exposed myself to the
rage of both civil and religious factions, they seemed to be
disarmed in my behalf of their wonted fury. My friends
never had occasion to vindicate any one circumstance of my
character and conduct ; not but that the zealots, we may well
suppose, would have been glad to invent and propagate any
story to my disadvantage but they could never find any which
they thought would wear the face of probability. I cannot say
there is no vanity in making this funeral oration of myself, but
I hope it is not a misplaced one ; and this is a matter of fact
which is easily cleared and ascertained."

Hume died in Edinburgh on the 25th of August,
1776, and, a few days later, his body, attended by
a great concourse of people, who seemed to have
anticipated for it the fate appropriate to the remains of wizards and necromancers, was deposited
in a spot selected by himself, in an old burial-
ground on the eastern slope of the Calton Hill.

From the summit of this hill, there is a prospect
unequalled by any to be seen from the midst of a
great city.   Westward lies the Forth, and beyond
it, dimly blue, the far away Highland hills ; east-
ward, rise the bold contours of Arthur's Seat and
the rugged crags of the Castle rock, with the gray
Old Town of Edinburgh ; while, far below, from a

maze of crowded thoroughfares, the hoarse murmur of the toil of a polity of energetic men is borne upon the ear. At times a man may be as solitary here as in a veritable wilderness; and may meditate undisturbedly upon the epitome of nature and of man—the kingdoms of this world—spread out before him.

Surely, there is a fitness in the choice of this last resting-place by the philosopher and historian, who saw so clearly that these two kingdoms form but one realm, governed by uniform laws and alike based on impenetrable darkness and eternal silence; and faithful to the last to that profound veracity which was the secret of his philosophic greatness, he ordered that the simple Roman tomb which marks his grave should bear no inscription but

DAVID HUME

BORN 1711.　　DIED 1776.

Leaving it to posterity to add the rest.

It was by the desire and at the suggestion of my friend, the Editor of this Series,[1] that I undertook to attempt to help posterity in the difficult business of knowing what to add to Hume's epitaph; and I might, with justice, throw upon him the responsibility of my apparent presumption in occupying a place among the men of

[1] *English Men of Letters.* Edited by John Morley.

letters, who are engaged with him, in their proper function of writing about English Men of Letters. That to which succeeding generations have made, are making, and will make, continual additions, however, is Hume's fame as a philosopher; and, though I know that my plea will add to my offence in some quarters, I must plead, in extenuation of my audacity, that philosophy lies in the province of science, and not in that of letters.

In dealing with Hume's Life, I have endeavoured, as far as possible, to make him speak for himself. If the extracts from his letters and essays which I have given do not sufficiently show what manner of man he was, I am sure that nothing I could say would make the case plainer. In the exposition of Hume's philosophy which follows, I have pursued the same plan, and I have applied myself to the task of selecting and arranging in systematic order, the passages which appeared to me to contain the clearest statements of Hume's opinions.

I should have been glad to be able to confine myself to this duty, and to limit my own comments to so much as was absolutely necessary to connect my excerpts. Here and there, however, it must be confessed that more is seen of my thread than of Hume's beads. My excuse must be an ineradicable tendency to try to make things clear; while, I may further hope, that there is nothing in what I may have said, which is inconsistent with the logical development of Hume's principles.

My authority for the facts of Hume's life is the admirable biography, published in 1846, by Mr. John Hill Burton. The edition of Hume's works from which all citations are made is that published by Black and Tait in Edinburgh, in 1826. In this edition, the Essays are reprinted from the edition of 1777, corrected by the author for the press a short time before his death. It is well printed in four handy volumes; and as my copy has long been in my possession, and bears marks of much reading, it would have been troublesome for me to refer to any other. But, for the convenience of those who possess some other edition, the following table of the contents of the edition of 1826, with the paging of the four volumes, is given :—

VOLUME III.

ESSAYS, MORAL AND POLITICAL, p. 3—p. 282

POLITICAL DISCOURSES, p. 285—p. 579

VOLUME IV.

AN INQUIRY CONCERNING THE HUMAN UNDERSTANDING,
p. 3—p. 233.

AN INQUIRY CONCERNING THE PRINCIPLES OF MORALS,
p. 237—p. 431.

THE NATURAL HISTORY OF RELIGION, p. 435—p. 513.

ADDITIONAL ESSAYS, p. 517—p. 577.

As the volume and the page of the volume are given in my references, it will be easy, by the help of this table, to learn where to look for any passage cited, in differently arranged editions.

PART II
HUME'S PHILOSOPHY

CHAPTER I

THE OBJECT AND SCOPE OF PHILOSOPHY

KANT has said that the business of philosophy is to answer three questions: What can I know? What ought I to do? and For what may I hope? But it is pretty plain that these three resolve themselves, in the long run, into the first. For rational expectation and moral action are alike based upon beliefs ; and a belief is void of justification, unless its subject-matter lies within the boundaries of possible knowledge, and unless its evidence satisfies the conditions which experience imposes as the guarantee of credibility.

Fundamentally, then, philosophy is the answer to the question, What can I know? and it is by applying itself to this problem, that philosophy is properly distinguished as a special department of scientific research. What is commonly called science, whether mathematical, physical, or biological, consists of the answers which mankind have been able to give to the inquiry, What

do I know ? They furnish us with the results of
the mental operations which constitute thinking;
while philosophy, in the stricter sense of the term,
inquires into the foundation of the first principles
which those operations assume or imply.

But though, by reason of the special purpose of
philosophy, its distinctness from other branches of
scientific investigation may be properly vindicated,
it is easy to see that, from the nature of its subject-
matter, it is intimately and, indeed, inseparably
connected with one branch of science. For it is
obviously impossible to answer the question, What
can we know ? unless, in the first place, there is a
clear understanding as to what is meant by knowledge ; and, having settled this point, the next
step is to inquire how we come by that which we
allow to be knowledge ; for, upon the reply,
turns the answer to the further question, whether,
from the nature of the case, there are limits to
the knowable or not. While, finally, inasmuch as
What can I know ? not only refers to knowledge
of the past or of the present, but to the confident
expectation which we call knowledge of the
future ; it is necessary to ask, further, what
justification can be alleged for trusting to the
guidance of our expectations in practical conduct.

It surely needs no argumentation to show, that
the first problem cannot be approached without
the examination of the contents of the mind ; and
the determination of how much of these contents

may be called knowledge. Nor can the second problem be dealt with in any other fashion; for it is only by the observation of the growth of knowledge that we can rationally hope to discover how knowledge grows. But the solution of the third problem simply involves the discussion of the data obtained by the investigation of the foregoing two.

Thus, in order to answer three out of the four subordinate questions into which What can I know? breaks up, we must have recourse to that investigation of mental phenomena, the results of which are embodied in the science of psychology.

Psychology is a part of the science of life or biology, which differs from the other branches of that science, merely in so far as it deals with the psychical, instead of the physical, phenomena of life.

As there is an anatomy of the body, so there is an anatomy of the mind; the psychologist dissects mental phenomena into elementary states of consciousness, as the anatomist resolves limbs into tissues, and tissues into cells. The one traces the development of complex organs from simple rudiments; the other follows the building up of complex conceptions out of simpler constituents of thought. As the physiologist inquires into the way in which the so-called "functions" of the body are performed, so the psychologist studies the so-called "faculties" of the mind. Even a cursory attention to the ways and works of the lower animals suggests a comparative anatomy and physiology of the mind; and the doctrine of evolution presses for application as much in the one field as in the other.

But there is more than a parallel, there is a close and intimate connection between psychology and physiology. No one doubts that, at any rate some mental states are dependent for their existence on the performance of the functions of particular bodily organs. There is no seeing without eyes, and no hearing without ears. If the origin of the contents of the mind is truly a philosophical problem, then the philosopher who attempts to deal with that problem, without acquainting himself with the physiology of sensation, has no more intelligent conception of his business than the physiologist, who thinks he can discuss locomotion, without an acquaintance with the principles of mechanics; or respiration, without some tincture of chemistry.

On whatever ground we term physiology, science, psychology is entitled to the same appellation; and the method of investigation which elucidates the true relations of the one set of phenomena will discover those of the other. Hence, as philosophy is, in great measure, the exponent of the logical consequences of certain data established by psychology; and as psychology itself differs from physical science only in the nature of its subject-matter, and not in its method of investigation, it would seem to be an obvious conclusion, that philosophers are likely to be successful in their inquiries, in proportion as they are familiar with the application of scientific method to less abstruse subjects; just as it seems to require no elaborate demonstration, that an astronomer, who wishes to comprehend the solar system, would do well to acquire a preliminary acquaintance with the elements of physics. And it is accordant with this presumption, that the men who have made the most important positive additions to philosophy, such as Descartes, Spinoza, and Kant, not to mention more recent examples, have been deeply imbued with the spirit of physical science; and, in some cases, such as those of Descartes and Kant, have been largely acquainted with its details. On the other hand, the founder of Positivism no less admirably illustrates the connection of scientific incapacity with philosophical incompetence. In truth, the laboratory is the fore-court of the temple of philosophy; and whoso has not offered sacrifices and undergone purification there, has little chance of admission into the sanctuary.

Obvious as these considerations may appear to be, it would be wrong to ignore the fact that their force is by no means universally admitted. On the contrary, the necessity for a proper psychological and physiological training to the student

of philosophy is denied, on the one hand, by the
" pure metaphysicians," who attempt to base the
theory of knowing upon supposed necessary and
universal truths, and assert that scientific observation is impossible unless such truths are already
known or implied : which, to those who are not
" pure metaphysicians," seems very much as if one
should say that the fall of a stone cannot be
observed, unless the law of gravitation is already
in the mind of the observer.

On the other hand, the Positivists, so far as
they accept the teachings of their master, roundly
assert, at any rate in words, that observation of
the mind is a thing inherently impossible in itself,
and that psychology is a chimera—a phantasm
generated by the fermentation of the dregs of
theology. Nevertheless, if M. Comte had been
asked what he meant by " physiologie cérébrale,"
except that which other people call " psychology " ;
and how he knew anything about the functions of
the brain, except by that very " observation
intérieure," which he declares to be an absurdity
—it seems probable that he would have found it
hard to escape the admission, that, in vilipending
psychology, he had been propounding solemn
nonsense.

It is assuredly one of Hume's greatest merits
that he clearly recognised the fact that philosophy is
based upon psychology ; and that the inquiry into
the contents and the operations of the mind must

be conducted upon the same principles as a physical investigation, if what he calls the "moral philosopher" would attain results of as firm and definite a character as those which reward the "natural philosopher."[1] The title of his first work, a "Treatise of Human Nature, being an Attempt to introduce the Experimental method of Reasoning into Moral Subjects," sufficiently indicates the point of view from which Hume regarded philosophical problems; and he tells us in the preface, that his object has been to promote the construction of a "science of man."

"'Tis evident that all the sciences have a relation, greater or less, to human nature; and that, however wide any of them may seem to run from it, they still return back by one passage or another. Even *Mathematics, Natural Philosophy,* and *Natural Religion* are in some measure dependent on the science of MAN; since they lie under the cognizance of men, and are judged of by their powers and qualities. 'Tis impossible to tell what changes and improvements we might make in these sciences were we thoroughly acquainted with the extent and force of human understanding, and could explain the nature of the ideas we employ and of the operations we perform in our reasonings . . . . To me it seems evident that the essence of mind being equally unknown to us with that of external bodies, it must be equally impossible to form any notion of its

---

[1] In a letter to Hutcheson (September 17th, 1739) Hume remarks:—"There are different ways of examining the mind as well as the body. One may consider it either as an anatomist or as a painter: either to discover its most secret springs and principles, or to describe the grace and beauty of its actions;" and he proceeds to justify his own mode of looking at the moral sentiments from the anatomist's point of view.

powers and qualities otherwise than from careful and exact experiments, and the observation of those particular effects which result from its different circumstances and situations. And though we must endeavour to render all our principles as universal as possible, by tracing up our experiments to the utmost, and explaining all effects from the simplest and fewest causes, 'tis still certain we cannot go beyond experience : and any hypothesis that pretends to discover the ultimate original qualities of human nature, ought at first to be rejected as pre-sumptuous and chimerical. . . . .

"But if this impossibility of explaining ultimate principles should be esteemed a defect in the science of man, I will ven-ture to affirm, that it is a defect common to it with all the sciences, and all the arts, in which we can employ ourselves, whether they be such as are cultivated in the schools of the philosophers, or practised in the shops of the meanest artizans. None of them can go beyond experience, or establish any principles which are not founded on that authority. Moral philosophy has, indeed, this peculiar disadvantage, which is not found in natural, that in collecting its experiments, it cannot make them purposely, with premeditation, and after such a manner as to satisfy itself concerning every particular diffi-culty which may arise. When I am at a loss to know the effects of one body upon another in any situation I need only put them in that situation, and observe what results from it. But should I endeavour to clear up in the same manner any [1] doubt in moral philosophy, by placing myself in the same case with that which I consider, 'tis evident this reflection and premeditation would so disturb the operation of my natural principles, as must render it impossible to form any just con-clusion from the phenomenon. We must, therefore, glean up our experiments in this science from a cautious observation of human life, and take them as they appear in the common course of the

---

[1] The manner in which Hume constantly refers to the results of the observation of the contents and the processes of his own mind clearly shows that he has here inadvertently overstated the case.

world, by men's behaviour in company, in affairs, and in their
pleasures. Where experiments of this kind are judiciously
collected and compared, we may hope to establish on them a
science which will not be inferior in certainty, and will be much
superior in utility, to any other of human comprehension."—(I.
pp. 7—11.)

All science starts with hypotheses—in other
words, with assumptions that are unproved, while
they may be, and often are, erroneous; but which
are better than nothing to the seeker after order
in the maze of phenomena. And the historical
progress of every science depends on the criticism
of hypotheses—on the gradual stripping off, that
is, of their untrue or superfluous parts—until
there remains only that exact verbal expression
of as much as we know of the fact, and no more,
which constitutes a perfect scientific theory.

Philosophy has followed the same course as
other branches of scientific investigation. The
memorable service rendered to the cause of sound
thinking by Descartes consisted in this: that he
laid   the   foundation   of   modern   philosophical
criticism   by   his   inquiry   into   the   nature   of
certainty. It is a clear result of the investigation
started by Descartes, that there is one thing of
which no doubt can be entertained, for he who
should pretend to doubt it would thereby prove
its   existence;   and   that   is   the   momentary
consciousness   we   call   a   present   thought   or
feeling; that is safe, even if all other kinds of

certainty are merely more or less probable inferences. Berkeley and Locke, each in his way, applied philosophical criticism in other directions ; but they always, at any rate professedly, followed the Cartesian maxim of admitting no propositions to be true but such as are clear, distinct, and evident, even while their arguments stripped off many a layer of hypothetical assumption which their great predecessor had left untouched. No one has more clearly stated the aims of the critical philosopher than Locke, in a passage of the famous "Essay concerning Human Understanding," which, perhaps, I ought to assume to be well known to all English readers, but which so probably is unknown to this full-crammed and much-examined generation that I venture to cite it :

"If by this inquiry into the nature of the understanding I can discover the powers thereof, how far they reach, to what things they are in any degree proportionate, and where they fail us, I suppose it may be of use to prevail with the busy mind of man to be more cautious in meddling with things exceeding his comprehension : to stop when it is at the utmost extent of its tether ; and to sit down in quiet ignorance of those things which, upon examination, are proved to be beyond the reach of our capacities. We should not then, perhaps, be so forward, out of an affectation of universal knowledge, to raise questions and perplex ourselves and others with disputes about things to which our understandings are not suited, and of which we cannot frame in our minds any clear and distinct perception, or whereof (as it has, perhaps, too often happened) we have not any notion at all . . . . Men may find matter sufficient to busy their heads and

employ their hands with variety, delight, and satisfaction, if
they will not boldly quarrel with their own constitution and
throw away the blessings their hands are filled with because
they are not big enough to grasp everything. We shall not
have much reason to complain of the narrowness of our minds,
if we will but employ them about what may be of use to us : for
of that they are very capable : and it will be an unpardonable,
as well as a childish peevishness, if we undervalue the advan-
tages of our knowledge, and neglect to improve it to the ends
for which it was given us, because there are some things that
are set out of reach of it. It will be no excuse to an idle and
untoward servant who would not attend to his business by
candlelight, to plead that he had not broad sunshine. The
candle that is set up in us shines bright enough for all our
purposes . . . . Our business here is not to know all
things, but those which concern our conduct."[1]

Hume develops the same fundamental conception in a somewhat different way, and with
a more definite indication of the practical benefits
which may be expected from a critical philosophy.
The first and second parts of the twelfth section
of the " Inquiry " are devoted to a condemnation
of excessive scepticism, or Pyrrhonism, with which
Hume couples a caricature of the Cartesian
doubt ; but, in the third part, a certain " mitigated
scepticism " is recommended and adopted, under
the title of " academical philosophy." After
pointing out that a knowledge of the infirmities
of the human understanding, even in its most perfect state, and when most accurate and cautious

[1] Locke, *An Essay concerning Human Understanding*, Book
I. chap i. §§ 4, 5, 6.

in its determinations, is the best check upon the tendency to dogmatism, Hume continues:—

"Another species of *mitigated* scepticism, which may be of advantage to mankind, and which may be the natural result of the PYRRHONIAN doubts and scruples, is the limitation of our inquiries to such subjects as are best adapted to the narrow capacity of human understanding. The *imagination* of man is naturally sublime, delighted with whatever is remote and extraordinary, and running, without control, into the most distant parts of space and time in order to avoid the objects which custom has rendered too familiar to it. A correct *judgment* observes a contrary method, and, avoiding all distant and high inquiries, confines itself to common life, and to such subjects as fall under daily practice and experience; leaving the more sublime topics to the embellishment of poets and orators, or to the arts of priests and politicians. To bring us to so salutary a determination, nothing can be more serviceable than to be once thoroughly convinced of the force of the PYRRHONIAN doubt, and of the impossibility that anything but the strong power of natural instinct could free us from it. Those who have a propensity to philosophy will still continue their researches; because they reflect, that, besides the immediate pleasure attending such an occupation, philosophical decisions are nothing but the reflections of common life, methodised and corrected. But they will never be tempted to go beyond common life, so long as they consider the imperfection of those faculties which they employ, their narrow reach and their inaccurate operations. While we cannot give a satisfactory reason why we believe, after a thousand experiments, that a stone will fall or fire burn; can we ever satisfy ourselves concerning any determination which we may form with regard to the origin of worlds and the situation of nature from and to eternity?" (IV. pp. 189—90.)

But further, it is the business of criticism not only to keep watch over the vagaries of philosophy, but to do the duty of police in the whole world of thought. Wherever it espies sophistry or superstition they are to be bidden to stand; nay, they are to be followed to their very dens and there apprehended and exterminated, as Othello smothered Desdemona, "else she'll betray more men."

Hume warms into eloquence as he sets forth the labours meet for the strength and the courage of the Hercules of "mitigated scepticism."

"Here, indeed, lies the justest and most plausible objection against a considerable part of metaphysics, that they are not properly a science, but arise either from the fruitless efforts of human vanity, which would penetrate into subjects utterly inaccessible to the understanding, or from the craft of popular superstitions, which, being unable to defend themselves on fair ground, raise these entangling brambles to cover and protect their weakness. Chased from the open country, these robbers fly into the forest, and lie in wait to break in upon every unguarded avenue of the mind and overwhelm it with religious fears and prejudices. The stoutest antagonist, if he remits his watch a moment, is oppressed; and many, through cowardice and folly, open the gates to the enemies, and willingly receive them with reverence and submission as their legal sovereigns.

"But is this a sufficient reason why philosophers should desist from such researches and leave superstition still in possession of her retreat? Is it not proper to draw an opposite conclusion, and perceive the necessity of carrying the war into the most secret recesses of the enemy? . . . . . The only method of freeing learning at once from these abstruse questions, is to inquire seriously into the nature of human understanding, and show, from an exact analysis of its powers and capacity, that it is by no means fitted for such remote and abstruse

subjects. We must submit to this fatigue, in order to live at
ease ever after ; and must cultivate true metaphysics with some
care, in order to destroy the false and adulterated."—(IV. pp.
10, 11.)

Near a century and a half has elapsed since
these brave words were shaped by David Hume's
pen ; and the business of carrying the war into
the enemy's camp has gone on but slowly. Like
other campaigns, it long languished for want of a
good base of operations. But since physical
science, in the course of the last fifty years, has
brought to the front an inexhaustible supply of
heavy artillery of a new pattern, warranted to
drive solid bolts of fact through the thickest
skulls, things are looking better ; though hardly
more than the first faint flutterings of the dawn
of the happy day, when superstition and false
metaphysics shall be no more and reasonable folks
may "live at ease," are as yet discernible by the
enfants perdus of the outposts.

If, in thus conceiving the object and the
limitations of philosophy, Hume shows himself
the spiritual child and continuator of the work of
Locke, he appears no less plainly as the parent of
Kant and as the protagonist of that more modern
way of thinking, which has been called "agnosti-
cism," from its profession of an incapacity to
discover the indispensable conditions of either
positive or negative knowledge, in many propositions, respecting which, not only the vulgar,

but philosophers of the more sanguine sort, revel in the luxury of unqualified assurance.

The aim of the " Kritik der reinen Vernunft " is essentially the same as that of the " Treatise of Human Nature," by which indeed Kant was led to develop that " critical philosophy " with which his name and fame are indissolubly bound up : and, if the details of Kant's criticism differ from those of Hume, they coincide with them in their main result, which is the limitation of all know-ledge of reality to the world of phenomena re-vealed to us by experience.

The philosopher of Königsberg epitomises the philosopher of Ninewells when he thus sums up the uses of philosophy :—

"'The greatest and perhaps the sole use of all philosophy of pure reason is, after all, merely negative, since it serves, not as an organon for the enlargement [of knowledge], but as a discip-line for its delimitation :' and instead of discovering truth, has only the modest merit of preventing error."[1]

---

[1] *Kritik der reinen Vernunft.*   Ed. Hartenstein, p. 256.

CHAPTER II

In the language of common life, the "mind" is spoken of as an entity, independent of the body, though resident in and closely connected with it, and endowed with numerous "faculties," such as sensibility, understanding, memory, volition, which stand in the same relation to the mind as the organs do to the body, and perform the functions of feeling, reasoning, remembering, and willing. Of these functions, some, such as sensation, are supposed to be merely passive—that is, they are called into existence by impressions, made upon the sensitive faculty by a material world of real objects, of which our sensations are supposed to give us pictures; others, such as the memory and the reasoning faculty, are considered to be partly passive and partly active; while volition is held to be potentially, if not always actually, a spontaneous activity.

The popular classification and terminology of
the phenomena of consciousness, however, are by
no means the first crude conceptions suggested by
common sense, but rather a legacy, and, in many
respects, a sufficiently *damnosa hæreditas*, of
ancient philosophy, more or less leavened by
theology; which has incorporated itself with the
common thought of later times, as the vices of the
aristocracy of one age become those of the mob in
the next. Very little attention to what passes in
the mind is sufficient to show, that these conceptions involve assumptions of an extremely
hypothetical character. And the first business
of the student of psychology is to get rid of such
prepossessions; to form conceptions of mental
phenomena as they are given us by observation,
without any hypothetical admixture, or with only
so much as is definitely recognised and held
subject to confirmation or otherwise; to classify
these phenomena according to their clearly
recognisable characters; and to adopt a nomenclature which suggests nothing beyond the results
of observation. Thus chastened, observation of
the mind makes us acquainted with nothing but
certain events, facts, or phenomena (whichever
name be preferred) which pass over the inward
field of view in rapid and, as it may appear on
careless inspection, in disorderly succession, like
the shifting patterns of a kaleidoscope. To all
these mental phenomena, or states of our

consciousness,[1] Descartes gave the name of
" thoughts," [2] while Locke and Berkeley termed
them " ideas." Hume, regarding this as an im-
proper use of the word " idea," for which he
proposes another employment, gives the general
name of " perceptions " to all states of consciousness. Thus, whatever other signification we may
see reason to attach to the word " mind," it is certain that it is a name which is employed to denote
a series of perceptions; just as the word " tune,"
whatever else it may mean, denotes, in the first
place, a succession of musical notes. Hume,
indeed, goes further than others when he says
that—

> " What we call a mind is nothing but a heap or collection of
> different perceptions, united together by certain relations, and
> supposed, though falsely, to be endowed with a perfect simplicity
> and identity."—(I. p. 268.)

With this " nothing but," however, he obviously
falls into the primal and perennial error of
philosophical speculators—dogmatising from negative arguments. He may be right or wrong; but

---

[1] " Consciousnesses " would be a better name, but it is
awkward. I have elsewhere proposed *psychoses* as a substantive
name for mental phenomena.

[2] As this has been denied, it may be as well to give
Descartes's words : Par le mot de penser, j'entends tout ce
que se fait dans nous de telle sorte que nous l'apercevons
immédiatement par nousmêmes : c'est pourquoi non-seulement
entendre, vouloir, imaginer, mais aussi sentir, c'est le même
chose ici que penser."—*Principes de Philosophie.* Ed. Cousin, 57.

"Toutes les propriétés que nous trouvons en la chose qui
pense ne sont que des façons différentes de penser."—*Ibid.* 96.

the most he, or anybody else, can prove in favour
of his conclusion is, that we know nothing more
of the mind than that it is a series of perceptions.
Whether there is something in the mind that
lies beyond the reach of observation; or whether
perceptions themselves are the products of something which can be observed and which is not
mind; are questions which can in nowise be
settled by direct observation. Elsewhere, the
objectionable hypothetical element of the definition of mind is less prominent :—

'The true idea of the human mind is to consider it as a
system of different perceptions, or different existences, which
are linked together by the relation of cause and effect, and
mutually produce, destroy, influence and modify each other. . . .
In this respect I cannot compare the soul more properly to
anything than a republic or commonwealth, in which the
several members are united by the reciprocal ties of government
and surbordination, and give rise to other persons who propagate the same republic in the incessant changes of its parts."—
(I. p. 331).

But, leaving the question of the proper definition of mind open for the present, it is further a
matter of direct observation, that, when we take
a general survey of all our perceptions or states of
consciousness, they naturally fall into sundry
groups or classes. Of these classes, two are
distinguished by Hume as of primary importance,
All " perceptions," he says, are either " *Impressions* " or " *Ideas.*"

Under " impressions " he includes " all our more

lively perceptions, when we hear, see, feel, love,
or will;" in other words, "all our sensations,
passions, and emotions, as they make their first
appearance in the soul" (I. p. 15).

"Ideas," on the other hand, are the faint images
of impressions in thinking and reasoning, or of
antecedent ideas.

Both impressions and ideas may be either
simple, when they are incapable of further
analysis, or *complex*, when they may be resolved
into simpler constituents. All simple ideas are
exact copies of impressions; but, in complex ideas,
the arrangement of simple constituents may be
different from that of the impressions of which
those simple ideas are copies.

Thus the colours red and blue and the odour of
a rose, are simple impressions; while the ideas of
blue, of red, and of rose-odour are simple copies of
these impressions. But a red rose gives us a
complex impression, capable of resolution into the
simple impressions of red colour, rose-scent, and
numerous others; and we may have a complex
idea, which is an accurate, though faint, copy of
this complex impression. Once in possession of
the ideas of a red rose and of the colour blue, we
may, in imagination, substitute blue for red; and
thus obtain a complex idea of a blue rose, which
is not an actual copy of any complex impression,
though all its elements are such copies.

Hume has been criticised for making the

distinction of impressions and ideas to depend upon their relative strength or vivacity. Yet it would be hard to point out any other character by which the things signified can be distinguished. Any one who has paid attention to the curious subject of what are called " subjective sensations " will be familiar with examples of the extreme difficulty which sometimes attends the discrimination of ideas of sensation from impressions of sensation, when the ideas are very vivid, or the impressions are faint. Who has not "fancied" he heard a noise ; or has not explained inattention to a real sound by saying, " I thought it was nothing but my fancy " ? Even healthy persons are much more liable to both visual and auditory spectra— that is, ideas of vision and sound so vivid that they are taken for new impressions—than is commonly supposed ; and, in some diseased states, ideas of sensible objects may assume all the vividness of reality.

If ideas are nothing but copies of impressions, arranged, either in the same order as that of the impressions from which they are derived, or in a different order, it follows that the ultimate analysis of the contents of the mind turns upon that of the impressions. According to Hume, these are of two kinds : either they are impressions of sensation, or they are impressions of reflection. The former are those afforded by the five senses, together with pleasure and pain. The latter are the passions or the emotions (which Hume employs as equivalent terms). Thus the elementary states of consciousness, the raw materials of knowledge, so to speak, are either sensations or emotions; and whatever we discover in the mind, beyond these elementary states of consciousness, results from the combinations and the metamorphoses which they undergo.

It is not a little strange that a thinker of Hume's capacity should have been satisfied with the results of a psychological analysis which regards some obvious compounds as elements, while it omits altogether a most important class of elementary states.

With respect to the· former point, Spinoza's masterly examination of the Passions in the third part of the "Ethics" should have been known to Hume.[1] But, if he had been acquainted with that wonderful piece of psychological anatomy, he would have learned that the emotions and passions are all complex states, arising from the close association of ideas of pleasure or pain with other ideas; and, indeed, without going to Spinoza, his own acute discussion of the passions leads to the same result,[2] and is wholly inconsistent

[1] On the whole, it is pleasant to find satisfactory evidence that Hume knew nothing of the works of Spinoza; for the invariably abusive manner in which he refers to that type of the philosophic hero is only to be excused, if it is to be excused, by sheer ignorance of his life and work.

[2] For example, in discussing pride and humility, Hume says :—
" According as our idea of ourselves is more or less advantageous,

with his classification of those mental states among the primary uncompounded materials of consciousness.

If Hume's "impressions of reflection" are excluded from among the primary elements of consciousness, nothing is left but the impressions afforded by the five senses, with pleasure and pain. Putting aside the muscular sense, which had not come into view in Hume's time, the questions arise whether these are all the simple undecomposable materials of thought? or whether others exist of which Hume takes no cognizance?

Kant answered the latter question in the affirmative, in the "Kritik der reinen Vernunft," and thereby made one of the greatest advances ever effected in philosophy; though it must be confessed that the German philosopher's exposition of his views is so perplexed in style, so burdened with the weight of a cumbrous and uncouth scholasticism, that it is easy to confound the unessential parts of his system with those

we feel either of these opposite affections, and are elated by pride or dejected with humility . . . when self enters not into the consideration there is no room either for pride or humility." That is, pride is pleasure, and humility is pain, associated with certain conceptions of one's self; or as Spinoza puts it :— "Superbia est de se præ amore sui plus justo sentire" ("amor" being "lætitia concomitante idea causæ externæ"); and "Humilitas est tristitia orta ex eo quod homo suam impotentiam sive imbecillitatem contemplatur."

which are of profound importance. His baggage
train is bigger than his army, and the student
who attacks him is too often led to suspect he has
won a position when he has only captured a mob
of useless camp-followers.

In his "Principles of Psychology," Mr. Herbert
Spencer appears to me to have brought out the
essential truth which underlies Kant's doctrine in
a far clearer manner than any one else; but, for
the purpose of the present summary view of
Hume's philosophy, it must suffice if I state the
matter in my own way, giving the broad outlines,
without entering into the details of a large and
difficult discussion.

When a red light flashes across the field of
vision, there arises in the mind an "impression of
sensation"—which we call red. It appears to me
that this sensation, red, is a something which may
exist altogether independently of any other impression, or idea, as an individual existence. It
is perfectly conceivable that a sentient being
should have no sense but vision, and that he
should have spent his existence in absolute darkness, with the exception of one solitary flash of
red light. That momentary illumination would
suffice to give him the impression under consideration. The whole content of his consciousness
might be that impression; and, if he were endowed with memory, its idea.

Such being the state of affairs, suppose a second flash of red light to follow the first. If there were no memory of the latter, the state of the mind on the second occasion would simply be a repetition of that which occurred before. There would be merely another impression.

But suppose memory to exist, and that an idea of the first impression is generated; then, if the supposed sentient being were like ourselves, there might arise in his mind two altogether new impressions. The one is the feeling of the *succession* of the two impressions, the other is the feeling of their *similarity*.

Yet a third case is conceivable. Suppose two flashes of red light to occur together, then a third feeling might arise which is neither succession nor similarity, but that which we call *co-existence*.

These feelings, or their contraries, are the foundation of everything that we call a relation. They are no more capable of being described than sensations are; and, as it appears to me, they are as little susceptible of analysis into simpler elements. Like simple tastes and smells, or feelings of pleasure and pain, they are ultimate irresolvable facts of conscious experience; and, if we follow the principle of Hume's nomenclature, they must be called *impressions of relation*. But it must be remembered, that they differ from the other impressions, in requiring the pre-existence of at least two of the latter. Though devoid of the slightest resemblance to the other impressions, they are, in a manner, generated by them. In fact, we may regard them as a kind of impressions of impressions; or as the sensations of an inner sense, which takes cognizance of the materials furnished to it by the outer senses.

Hume failed as completely as his predecessors had done to recognise the elementary character of impressions of relation; and, when he discusses relations, he falls into a chaos of confusion and self-contradiction.

In the "Treatise," for example, (Book I., § iv.) resemblance, contiguity in time and space, and cause and effect, are said to be the "uniting principles among ideas," "the bond of union" or "associating quality by which one idea naturally introduces another." Hume affirms that—

"These qualities produce an association among ideas, and upon the appearance of one idea naturally introduce another." They are "the principles of union or cohesion among our simple ideas, and, in the imagination, supply the place of that inseparable connection by which they are united in our memory. Here is a kind of *attraction*, which, in the mental world, will be found to have as extraordinary effects as in the natural, and to show itself in as many and as various forms. Its effects are everywhere conspicuous; but, as to its causes they are mostly unknown, and must be resolved into *original* qualities of human nature, which I pretend not to explain."—(I. p. 29.)

And at the end of this section Hume goes on to say—

"Amongst the effects of this union or association of ideas there are none more remarkable than those complex ideas which are the common subjects of our thought and reasoning, and generally arise from some principle of union among our simple ideas. These complex ideas may be resolved into *relations, modes,* and *substances.*"—(*Ibid.*)

In the next section, which is devoted to *Relations,* they are spoken of as qualities "by which two ideas are connected together in the imagination," or "which make objects admit of comparison," and seven kinds of relation are enumerated, namely, *resemblance, identity, space and time, quantity or number, degrees of quality, contrariety,* and *cause and effect.*

To the reader of Hume, whose conceptions are usually so clear, definite, and consistent, it is as unsatisfactory as it is surprising to meet with so much questionable and obscure phraseology in a small space. One and the same thing, for example, resemblance, is first called a "quality of an idea," and secondly a "complex idea." Surely it cannot be both. Ideas which have the qualities of "resemblance, contiguity, and cause and effect," are said to "attract one another" (save the mark !), and so become associated; though, in a subsequent part of the "Treatise," Hume's great effort is to prove that the relation of cause and effect is a particular case of the

G 2

process of association; that is to say, is a result
of the process of which it is supposed to be the
cause. Moreover, since, as Hume is never weary
of reminding his readers, there is nothing in ideas
save copies of impressions, the qualities of resemblance, contiguity, and so on, in the idea, must
have existed in the impression of which that idea
is a copy; and therefore they must be either
sensations or emotions—from both of which
classes they are excluded.

In fact, in one place, Hume himself has an
insight into the real nature of relations. Speaking
of equality, in the sense of a relation of quantity,
he says—

"Since equality is a relation, it is not, strictly speaking, a
property in the figures themselves, but arises merely from the
comparison which the mind makes between them."—(I. p.
70.)

That is to say, when two impressions of equal
figures are present, there arises in the mind a
tertium quid, which is the perception of equality.
On his own principles, Hume should therefore
have placed this "perception" among the ideas of
reflection. However, as we have seen, he expressly excludes everything but the emotions and
the passions from this group.

It is necessary therefore to amend Hume's
primary "geography of the mind" by the exci-
sion of one territory and the addition of another:

and the elementary states of consciousness under consideration will stand thus :—

A. IMPRESSIONS.
    A. Sensations of
       a. Smell.
       b. Taste.
       c. Hearing.
       d. Sight.
       e. Touch.
       f. Resistance (the muscular sense).
    B. Pleasure and Pain.
    C. Relations.
       a. Co-existence.
       b. Succession.
       c. Similarity and dissimilarity.

B. IDEAS.
Copies, or reproductions in memory, of the foregoing.

And now the question arises, whether any, and if so what, portion of these contents of the mind are to be termed " knowledge ? "

According to Locke, " Knowledge is the perception of the agreement or disagreement of two ideas ; " and Hume, though he does not say so in so many words, tacitly accepts the definition. It follows, that neither simple sensation, nor simple emotion, constitutes knowledge ; but that, when impressions of relation are added to these impressions, or their ideas, knowledge arises; and that all knowledge is the knowledge of likenesses and unlikenesses, co-existences and successions.

It really matters very little in what sense terms are used, so long as the same meaning is always rigidly attached to them; and, therefore, it .is hardly worth while to quarrel with this generally accepted, though very arbitrary, limitation of the signification of "knowledge." But, on the face of the matter, it is not obvious why the impression we call a relation should have a better claim to the title of knowledge, than that which we call a sensation or an emotion; and the restriction has this unfortunate result, that it excludes all the most intense states of consciousness from any claim to the title of "knowledge."

For example, on this view, pain, so violent and absorbing as to exclude all other forms of consciousness, is not knowledge; but becomes a part of knowledge the moment we think of it in relation to another pain, or to some other mental phenomenon. Surely this is somewhat inconvenient, for there is only a verbal difference between having a sensation and knowing one has it: they are simply two phrases for the same mental state.

But the "pure metaphysicians" make great capital out of the ambiguity. For, starting with the assumption that all knowledge is the perception of relations, and finding themselves like mere common-sense folks, very much disposed to call sensation knowledge, they at once gratify that disposition and save their consistency, by declaring that even the simplest act of sensation contains, two terms and a relation—the sensitive subject the sensigenous object, and that masterful entity, the Ego. From which great triad, as from a gnostic Trinity, emanates an endless procession of other logical shadows and all the *Fata Morgana* of philosophical dreamland.

CHAPTER III

ADMITTING that the sensations, the feelings of pleasure and pain, and those of relation, are the primary irresolvable states of consciousness, two further lines of investigation present themselves. The one leads us to seek the origin of these "impressions:" the other, to inquire into the nature of the steps by which they become metamorphosed into those compound states of consciousness, which so largely enter into our ordinary trains of thought.

With respect to the origin of impressions of sensation, Hume is not quite consistent with himself. In one place (I. p. 117) he says, that it is impossible to decide "whether they arise immediately from the object, or are produced by the creative power of the mind, or are derived from the Author of our being," thereby implying that realism and idealism are equally probable hypotheses. But, in fact, after the demonstration by Descartes, that

the immediate antecedents of sensations are
changes in the nervous system, with which our
feelings have no sort of resemblance, the hypothesis that sensations " arise immediately from
the object " was out of court; and that Hume fully
admitted the Cartesian doctrine is apparent when
he says (I. p. 272) :—

" All our perceptions are dependent on our organs and the
disposition of our nerves and animal spirits."

And again, though in relation to another question,
he observes :—

" There are three different kinds of impressions conveyed
by the senses. The first are those of the figure, bulk, motion,
and solidity of bodies. The second those of colours, tastes,
smells, sounds, heat, and cold. The third are the pains and
pleasures that arise from the application of objects to our
bodies, as by the cutting of our flesh with steel, and such like.
Both philosophers and the vulgar suppose the first of these to
have a distinct continued existence. The vulgar only regard
the second as on the same footing. Both philosophers and
the vulgar again esteem the third to be merely perceptions,
and consequently interrupted and dependent beings.

" Now 'tis evident that, whatever may be our philosophical
opinion, colour, sounds, heat, and cold, as far as appears to
the senses, exist after the same manner with motion and
solidity ; and that the difference we make between them, in
this respect, arises not from the mere perception. So strong
is the prejudice for the distinct continued existence of the
former qualities, that when the contrary opinion is advanced
by modern philosophers, people imagine they can almost
refute it from their reason and experience, and that their very
senses contradict this philosophy. 'Tis also evident that
colours, sounds, &c., are originally on the same footing with
the pain that arises from steel, and pleasure that proceeds from

a fire ; and that the difference betwixt them is founded neither
on perception nor reason, but on the imagination. For as they
are confessed to be, both of them, nothing but perceptions
arising from the particular configurations and motions of the
parts of the body, wherein possibly can their difference consist ?
Upon the whole then, we may conclude that, as far as the senses
are judges all perceptions are the same in the manner of their
existence."—(I. p. 250, 251.)

The last words of this passage are as much
Berkeley's as Hume's. But, instead of following
Berkeley in his deductions from the position thus
laid down, Hume, as the preceding citation
shows, fully adopted the conclusion to which all
that we know of psychological physiology tends,
that the origin of the elements of consciousness,
no less than that of all its other states, is to be
sought in bodily changes, the seat of which can
only be placed in the brain. And, as Locke had
already done with less effect, he states and refutes
the arguments commonly brought against the
possibility of a causal connection between the
modes of motion of the cerebral substance and
states of consciousness, with great clearness :— ·

"From these hypotheses concerning the *substance* and *local
conjunction* of our perceptions we may pass to another, which
is more intelligible than the former, and more important than
the latter, viz. concerning the *cause* of our perceptions. Matter
and motion, 'tis commonly said in the schools, however varied,
are still matter and motion, and produce only a difference in
the position and situation of objects. Divide a body as often as
you please, 'tis still body. Place it in any figure, nothing ever
results but figure, or the relation of parts. Move it in any

manner, you still find motion or a change of relation. 'Tis absurd to imagine that motion in a circle, for instance, should be nothing but merely motion in a circle ; while motion in another direction, as in an ellipse, should also be a passion or moral reflection ; that the shocking of two globular particles should become a sensation of pain, and that the meeting of the triangular ones should afford a pleasure. Now as these different shocks and variations and mixtures are the only changes of which matter is susceptible, and as these never afford us any idea of thought or perception, 'tis concluded to be impossible, that thought can ever be caused by matter._

"Few have been able to withstand the seeming evidence of this argument ; and yet nothing in the world is more easy than to refute it. We need only reflect upon what has been proved at large, that we are never sensible of any connection between causes and effects, and that 'tis only by our experience of their constant conjunction we can arrive at any knowledge of this relation. Now, as all objects which are not contrary are susceptible of a constant conjunction, and as no real objects are contrary, I have inferred from these principles (Part III. § 15) that, to consider the matter *a priori*, anything may produce anything, and that we shall never discover a reason why any object may or may not be the cause of any other, however great, or however little, the resemblance may be betwixt them. This evidently destroys the precedent reasoning, concerning the cause of thought or perception. For though there appear no manner of connection betwixt motion and thought, the case is the same with all other causes and effects. Place one body of a pound weight on one end of a lever, and another body of the same weight on the other end ; you will never find in these bodies any principle of motion dependent on their distance from the centre, more than of thought and perception. If you pretend, therefore, to prove, *a priori*, that such a position of bodies can never cause thought, because, turn it which way you will, it is nothing but a position of bodies : you must, by the same course of reasoning, conclude that it can never produce motion, since there is no more apparent connection in the one than in the other.

But, as this latter conclusion is contrary to evident experience, and as 'tis possible we may have a like experience in the operations of the mind, and may perceive a constant conjunction of thought and motion, you reason too hastily when, from the mere consideration of the ideas, you conclude that 'tis impossible motion can ever produce thought, or a different position of parts give rise to a different passion or reflection. Nay, 'tis not only possible we may have such an experience, but 'tis certain we have it ; since every one may perceive that the different dispositions of his body change his thoughts and sentiments. And should it be said that this depends on the union of soul and body, I would answer, that we must separate the question concerning the substance of the mind from that concerning the cause of its thought ; and that, confining ourselves to the latter question, we find, by the comparing their ideas, that thought and motion are different from each other, and by experience, that they are constantly united ; which, being all the circumstances that enter into the idea of cause and effect, when applied to the operations of matter, we may certainly conclude that motion may be, and actually is, the cause of thought and perception."—(I. pp. 314—316.)

The upshot of all this is, that the " collection of perceptions," which constitutes the mind, is really a system of effects, the causes of which are to be sought in antecedent changes of the matter of the brain, just as the " collection of motions," which we call flying, is a system of effects, the causes of which are to be sought in the modes of motion of the matter of the muscles of the wings.

Hume, however, treats of this important topic only incidentally. He seems to have had very little acquaintance even with such physiology as was current in his time. At least, the only passage of his works, bearing on this subject, with which I

am acquainted, contains nothing but a very odd
version of the physiological views of Descartes :—

"When I received the relations of *resemblance, contiguity,*
and *causation,* as principles of union among ideas, without
examining into their causes, 'twas more in prosecution of my
first maxim, that we must in the end rest contented with experience, than for want of something specious and plausible
which I might have displayed on that subject. 'Twould have
been easy to have made an imaginary dissection of the brain,
and have shown why, upon our conception of any idea, the
animal spirits run into all the contiguous traces and rouse up
the other ideas that are related to it. But though I have
neglected any advantage which I might have drawn from this
topic in explaining the relations of ideas, I am afraid I must
here have recourse to it, in order to account for the mistakes
that arise from these relations. I shall therefore observe, that
as the mind is endowed with the power of exciting any idea it
pleases ; whenever it despatches the spirits into that region of
the brain in which the idea is placed ; these spirits always
excite the idea, when they run precisely into the proper traces
and rummage that cell which belongs to the idea. But as their
motion is seldom direct, and naturally turns a little to the one
side or to the other ; for this reason the animal spirits, falling
into the contiguous traces, present other related ideas, in lieu of
that which the mind desired at first to survey. This change we
are not always sensible of ; but continuing still the same train
of thought, make use of the related idea which is presented to
us and employ it in our reasonings, as if it were the same with
what we demanded. This is the cause of many mistakes
and sophisms in philosophy ; as will naturally be imagined,
and as it would be easy to show, if there was occasion."—(I.
p. 88.)

Perhaps it is as well for Hume's fame that the
occasion for further physiological speculations of
this sort did not arise. But, while admitting the

crudity of his notions and the strangeness of the
language in which they are couched, it must in
justice be remembered, that what are now known
as the elements of the physiology of the nervous
system were hardly dreamed of in the first half of
the eighteenth century ; and, as a further set off
to Hume's credit, it must be noted that he grasped
the fundamental truth, that the key to the comprehension of mental operations lies in the study
of the molecular changes of the nervous apparatus
by which they are originated.

Surely no one who is cognisant of the facts of
the case, nowadays, doubts that the roots of
psychology lie in the physiology of the nervous
system. What we call the operations of the mind
are functions of the brain, and the materials of
consciousness are products of cerebral activity.
Cabanis may have made use of crude and misleading phraseology when he said that the brain
secretes thought as the liver secretes bile; but
the conception which that much-abused phrase
embodies is, nevertheless, far more consistent
with fact than, the popular notion that the mind
is a metaphysical entity seated in the head, but as
independent of the brain as a telegraph operator
is of his instrument.

It is hardly necessary to point out that the
doctrine just laid down is what is commonly
called materialism. In fact, I am not sure that
the adjective "crass," which appears to have a

special charm for rhetorical sciolists, would not be applied to it. But it is, nevertheless, true that the doctrine contains nothing inconsistent with the purest idealism. For, as Hume remarks (as indeed Descartes had observed long before) :—

"'Tis not our body we perceive when we regard our limbs and members, but certain impressions which enter by the senses ; so that the ascribing a real and corporeal existence to these impressions, or to their objects, is an act of the mind as difficult to explain as that [the external existence of objects] which we examine at present."—(I. p. 249.)

Therefore, if we analyse the proposition that all mental phenomena are the effects or products of material phenomena, all that it means amounts to this : that whenever those states of consciousness which we call sensation, or emotion, or thought, come into existence, complete investigation will show good reason for the belief that they are preceded by those other phenomena of consciousness to which we give the names of matter and motion. All material changes appear, in the long run, to be modes of motion ; but our knowledge of motion is nothing but that of a change in the place and order of our sensations ; just as our knowledge of matter is restricted to those feelings of which we assume it to be the cause.

It has already been pointed out, that Hume must have admitted, and in fact does admit, the possibility that the mind is a Leibnitzian monad, or a Fichtean world-generating Ego, the universe of things being merely the picture produced by the evolution of the phenomena of consciousness. For any demonstration that can be given to the contrary effect, the "collection of perceptions" which makes up our consciousness may be an orderly phantasmagoria generated by the Ego, unfolding its successive scenes on the background of the abyss of nothingness; as a firework, which is but cunningly arranged combustibles, grows from a spark into a coruscation, and from a coruscation into figures, and words, and cascades of devouring fire, and then vanishes into the darkness of the night.

On the other hand, it must no less readily be allowed that, for anything that can be proved to the contrary, there may be a real something which is the cause of all our impressions; that sensations, though not likenesses, are symbols of that something; and that the part of that something, which we call the nervous system, is an apparatus for supplying us with a sort of algebra of fact, based on those symbols. A brain may be the machinery by which the material universe becomes conscious of itself. But it is important to notice that, even if this conception of the universe and of the relation of consciousness to its other components should be true, we should, nevertheless, be still bound by the limits of thought, still unable to refute the arguments of pure idealism. The more completely the materialistic position is admitted, the easier is it to show that the idealistic position is unassailable, if the idealist confines himself within the limits of positive knowledge.

Hume deals with the questions whether all our ideas are derived from experience, or whether, on the contrary, more or fewer of them are innate, which so much exercised the mind of Locke, after a somewhat summary fashion, in a note to the second section of the " Inquiry " :—

" It is probable that no more was meant by those who denied innate ideas, than that all ideas were copies of our impressions ; though it must be confessed that the terms which they employed were not chosen with such caution, nor so exactly defined, as to prevent all mistakes about their doctrine. For what is meant by *innate?* If innate be equivalent to natural, then all the perceptions and ideas of the mind must be allowed to be innate or natural, in whatever sense we take the latter word, whether in opposition to what is uncommon, artificial, or miraculous. If by innate be meant contemporary with our birth, the dispute seems to be frivolous ; nor is it worth while to inquire at what time thinking begins, whether before, at, or after our birth. Again, the word *idea* seems to be commonly taken in a very loose sense by Locke and others, as standing for any of our perceptions, our sensations and passions, as well as thoughts. Now in this sense I should desire to know what can be meant by asserting that self-love, or resentment of injuries, or the passion between the sexes is not innate ?

" But admitting these terms, *impressions* and *ideas,* in the sense above explained, and understanding by *innate* what is original or copied from no precedent perception, then we may assert that all our impressions are innate, and our ideas not innate."

It would seem that Hume did not think it worth while to acquire a comprehension of the

real points at issue in the controversy which he
thus carelessly dismisses.

Yet Descartes has defined what he means by
innate ideas with so much precision, that miscon-
ception ought to have been impossible. He says
that, when he speaks of an idea being "innate,"
he means that it exists potentially in the mind,
before it is actually called into existence by whatever is its appropriate exciting cause.

"I have never either thought or said," he writes, "that the
mind has any need of innate ideas [*idées naturelles*] which are
anything distinct from its faculty of thinking. But it is true
that observing that there are certain thoughts which arise
neither from external objects nor from the determination of my
will, but only from my faculty of thinking; in order to mark
the difference between the ideas or the notions which are the
forms of these thoughts, and to distinguish them from the
others, which may be called extraneous or voluntary, I have
called them innate. But I have used this term in the same
sense as when we say that generosity is innate in certain
families; or that certain maladies, such as gout or gravel, are
innate in others; not that children born in these families are
troubled with such diseases in their mother's womb; but
because they are born with the disposition or the faculty of
contracting them." [1]

His troublesome disciple, Regius, having asserted
that all our ideas come from observation or tradition, Descartes remarks :—

"So thoroughly erroneous is this assertion, that whoever has
a proper comprehension of the action of our senses, and under-

---

[1] Remarques de Rene Descartes sur un certain placard im-
prime aux Pays Bas vers la fin de l'annee, 1647.—Descartes,
Œuvres. Ed. Cousin, x. p. 71.

stands precisely the nature of that which is transmitted by them to our thinking faculty, will rather affirm that no ideas of things, such as are formed in thought, are brought to us by the senses, so that there is nothing in our ideas which is other than innate in the mind (*naturel a l esprit*), or in the faculty of thinking, if only certain circumstances are excepted, which belong only to experience. For example, it is experience alone which causes us to judge that such and such ideas, now present in our minds, are related to certain things which are external to us ; not, in truth, that they have been sent into our mind by these things, such as they are, by the organs of the senses ; but because these organs have transmitted something which has occasioned the mind, in virtue of its innate power, to form them at this time rather than at another. . . . .

"Nothing passes from external objects to the soul except certain motions of matter (*mouvemens corporels*), but neither these motions, nor the figures which they produce, are conceived by us as they exist in the sensory organs, as I have fully explained in my ' Dioptrics ' ; whence it follows that even the ideas of motion and of figures are innate (*naturellement en nous*). And, à fortiori, the ideas of pain, of colours, of sounds, and of all similar things must be innate, in order that the mind may represent them to itself, on the occasion of certain motions of matter with which they have no resemblance."

Whoever denies what is, in fact, an inconceivable proposition, that sensations pass, as such, from the external world into the mind, must admit the conclusion here laid down by Descartes, that, strictly speaking, sensations, and a fortiori, all the other contents of the mind, are innate. Or, to state the matter in accordance with the views previously expounded, that they are products of the inherent properties of the thinking organ, in which they lie potentially, before they are called into existence by their appropriate causes.

But if all the contents of the mind are innate, what is meant by experience?

It is the conversion, by unknown causes, of these innate potentialities into actual existences. The organ of thought, prior to experience, may be compared to an untouched piano, in which it may be properly said that music is innate, inasmuch as its mechanism contains, potentially, so many octaves of musical notes. The unknown cause of sensation which Descartes calls the "je ne sais quoi dans les objets" or "choses telles qu'elles sont," and Kant the "Noumenon" or "Ding an sich," is represented by the musician; who, by touching the keys, converts the potentiality of the mechanism into actual sounds. A note so produced is the equivalent of a single experience.

All the melodies and harmonies that proceed from the piano depend upon the action of the musician upon the keys. There is no internal mechanism which, when certain keys are struck, gives rise to an accompaniment of which the musician is only indirectly the cause. According to Descartes, however—and this is what is generally fixed upon as the essence of his doctrine of innate ideas—the mind possesses such an internal mechanism, by which certain classes of thoughts are generated, on the occasion of certain experiences. Such thoughts are innate, just as sensations are innate; they are not copies of sensations, any more than sensations are copies of motions; they are invariably generated in the mind, when certain experiences arise in it, just as sensations are invariably generated when certain bodily motions take place; they are universal, inasmuch as they arise under the same conditions in all men; they are necessary, because their genesis under these conditions is invariable. These innate thoughts are what Descartes terms " verites " or truths: that is beliefs—and his notions respecting them are plainly set forth in a passage of the " Principes."

"Thus far I have discussed that which we know as things: it remains that I should speak of that which we know as truths. For example, when we think that it is impossible to make anything out of nothing, we do not imagine that this proposition is a thing which exists, or a property of something, but we take it for a certain eternal truth, which has its seat in the mind (*pensée*), and is called a common notion or an axiom. Similarly, when we affirm that it is impossible that one and the same thing should exist and not exist at the same time; that that which has been created should not have been created; that he who thinks must exist while he thinks; and a number of other like propositions; these are only truths, and not things which exist outside our thoughts. And there is such a number of these that it would be wearisome to enumerate them: nor is it necessary to do so, because we cannot fail to know them when the occasion of thinking about them presents itself, and we are not blinded by any prejudices."

It would appear that Locke was not more familiar with Descartes' writings than Hume seems to have been; for, viewed in relation to the passages just cited, the arguments adduced in his famous polemic against innate ideas are totally irrelevant.

It has been shown that Hume practically, if not in so many words, admits the justice of Descartes' assertion that, strictly speaking, sensations are innate; that is to say, that they are the product of the reaction of the organ of the mind on the stimulus of an "unknown cause," which is Descartes' "je ne sais quoi." Therefore, the difference between Descartes' opinion and that of Hume resolves itself into this : Given sensation-experiences, can all the contents of consciousness be derived from the collocation and metamorphosis of these experiences? Or, are new elements of consciousness, products of an innate potentiality distinct from sensibility, added to these? Hume affirms the former position, Descartes the latter. If the analysis of the phenomena of consciousness given in the preceding pages is correct, Hume is in error; while the father of modern philosophy had a truer insight, though he overstated the case. For want of sufficiently searching psychological investigations, Descartes was led to suppose that innumerable ideas, the evolution of which in the course of experience can be demonstrated, were direct or innate products of the thinking faculty.

As has been already pointed out, it is the great merit of Kant that he started afresh on the track indicated by Descartes, and steadily upheld the doctrine of the existence of elements of consciousness, which are neither sense-experiences nor any modifications of them. We may demur to the expression that space and time are forms of sensory intuition ; but it imperfectly represents the great fact that co-existence and succession are mental phenomena not given in the mere sense experience.[1]

[1] "Wir konnen uns keinen Gegenstand denken, ohne durch Kategorien ; wir können keinen gedachten Gegenstand erkennen, ohne durch Anschauungen, die jenen Begriffen entsprechen. Nun sind alle unsere Anschauungen sinnlich, und diese Erkenntniss, so fern der Gegenstand derselben gegeben ist, ist empirisch. Empirische Erkenntniss aber ist Erfahrung. Folglich ist uns keine Erkenntniss *a priori* moglich, als lediglich von Gegenstanden moglicher Erfahrung."

"Aber diese Erkenntniss, die bloss auf Gegenstände der Erfahrung eingeschränkt ist, ist darum nicht alle von der Erfahrung entlehnt, sondern was sowohl die reinen Anschauungen, als die reinen Verstandesbegriffe betrifft, so sind sie Elemente der Erkenntniss die in uns *a priori* angetroffen werden."—*Kritik der reinen Vernunft. Elementarlehre*, p. 135. Without a glossary explanatory of Kant's terminology, this passage would be hardly intelligible in a translation ; but it may be paraphrased thus : All knowledge is founded upon experiences of sensation, but it is not all derived from those experiences ; inasmuch as the impressions of relation ("reine Anschauungen" ; "reine Verstandesbegriffe") have a potential or *a priori* existence in us, and by their addition to sense-experiences, constitute knowledge.

CHAPTER IV

THE CLASSIFICATION AND THE NOMENCLATURE OF MENTAL OPERATIONS

IF, as has been set forth in the preceding chapter, all mental states are effects of physical causes, it follows that what are called mental faculties and operations are, properly speaking, cerebral functions, allotted to definite, though not yet precisely assignable, parts of the brain.

These functions appear to be reducible to three groups, namely : Sensation, Correlation, and Ideation.

The organs of the functions of sensation and correlation are those portions of the cerebral substance, the molecular changes of which give rise to impressions of sensation and impressions of relation.

The changes in the nervous matter which bring about the effects which we call its functions, follow upon some kind of stimulus, and rapidly reaching their maximum, as rapidly die away   The effect of the irritation of a nerve-fibre on the cerebral substance with which it is connected may be compared to the pulling of a long bell-wire. The impulse takes a little time to reach the bell; the bell rings and then becomes quiescent, until another pull is given. So, in the brain, every sensation is the ring of a cerebral particle, the effect of a momentary impulse sent along a nerve-fibre.

If there were a complete likeness between the two terms of this very rough and ready comparison, it is obvious that there could be no such thing as memory. A bell records no audible sign of having been rung five minutes ago, and the activity of a sensigenous cerebral particle might similarly leave no trace. Under these circumstances, again, it would seem that the only impressions of relation which could arise would be those of co-existence and of similarity. For succession implies memory of an antecedent state.[1]

But the special peculiarity of the cerebral apparatus is, that any given function which has once been performed is very easily set a-going again, by causes more or less different from those to which it owed its origin. Of the mechanism of this generation of images of impressions or ideas (in Hume's sense), which may be termed *Ideation*, we know nothing at present, though the fact and its results are familiar enough.

---

[1] It is not worth while, for the present purpose, to consider whether, as all nervous action occupies a sensible time, the duration of one impression might not overlap that of the impression which follows it, in the case supposed.

During our waking, and many of our sleeping, hours, in fact, the function of ideation is in continual, if not continuous, activity. Trains of thought, as we call them, succeed one another without intermission, even when the starting of new trains by fresh sense-impressions is as far as possible prevented. The rapidity and the intensity of this ideational process are obviously dependent upon physiological conditions. The widest differences in these respects are constitutional in men of different temperaments; and are observable in oneself, under varying conditions of hunger and repletion, fatigue and freshness, calmness and emotional excitement. The influence of diet on dreams; of stimulants upon the fulness and the velocity of the stream of thought; the delirious phantasms generated by disease, by hashish, or by alcohol; will occur to every one as examples of the marvellous sensitiveness of the apparatus of ideation to purely physical influences.

The succession of mental states in ideation is not fortuitous, but follows the law of association, which may be stated thus: that every idea tends to be followed by some other idea which is associated with the first, or its impression, by a relation of succession, of contiguity, or of likeness.

Thus the idea of the word horse just now presented itself to my mind, and was followed in quick succession by the ideas of four legs, hoofs, teeth, rider, saddle, racing, cheating; all of which ideas are connected in my experience with the impression, or the idea, of a horse and with one another, by the relations of contiguity and succession. No great attention to what passes in the mind is needful to prove that our trains of thought are neither to be arrested, nor even permanently controlled, by our desires or emotions. Nevertheless they are largely influenced by them. In the presence of a strong desire, or emotion, the stream of thought no longer flows on in a straight course, but seems, as it were, to eddy round the idea of that which is the object of the emotion. Every one who has " eaten his bread in sorrow " knows how strangely the current of ideas whirls about the conception of the object of regret or remorse as a centre; every now and then, indeed, breaking away into the new tracts suggested by passing associations, but still returning to the central thought. Few can have been so happy as to have escaped the social bore, whose pet notion is certain to crop up whatever topic is started; while the fixed idea of the monomaniac is but the extreme form of the same phenomenon.

And as, on the one hand, it is so hard to drive away the thought we would fain be rid of; so, upon the other, the pleasant imaginations which we would so gladly retain are, sooner or later, jostled away by the crowd of claimants for birth into the world of consciousness; which hover as a sort of psychical possibilities, or inverse ghosts, the bodily presentments of spiritual phenomena to be, in the limbo of the brain. In that form of desire which is called "attention," the train of thought, held fast, for a time, in the desired direction, seems ever striving to get on to another line—and the junctions and sidings are so multitudinous!

The constitutents of trains of ideas may be grouped in various ways.

Hume says:—

"We find, by experience, that when any impression has been present in the mind, it again makes its appearance there as an idea, and this it may do in two different ways : either when, on its new appearance, it retains a considerable degree of its first vivacity, and is somewhat intermediate between an impression and an idea ; or when it entirely loses that vivacity, and is a perfect idea. The faculty by which we repeat our impressions in the first manner, is called the *memory*, and the other the *imagination*."—(I. pp. 23, 24.)

And he considers that the only difference between ideas of imagination and those of memory, except the superior vivacity of the latter, lies in the fact that those of memory preserve the original order of the impressions from which they are derived, while the imagination "is free to transpose and change its ideas."

The latter statement of the difference between memory and imagination is less open to cavil than the former, though by no means unassailable.

The special characteristic of a memory surely is not its vividness; but that it is a complex idea, in

which the idea of that which is remembered is
related by co-existence with other ideas, and by
antecedence with present impressions.

If I say I remember A. B., the chance acquaintance of ten years ago, it is not because my idea of
A. B. is very vivid—on the contrary, it is extremely
faint—but because that idea is associated with
ideas of impressions co-existent with those which
I call A. B. ; and that all these are at the end of
the long series of ideas, which represent that
much past time.   In truth I have a much more
vivid idea of Mr. Pickwick, or of Colonel New-
come, than I have of A. B.; but, associated
with the ideas of these persons, I have no idea
of their having ever been derived from the world
of impressions ; and so they are relegated to the
world of imagination.   On the other hand, the
characteristic of an imagination may properly be
said to lie not in its intensity, but in the fact, that
as Hume puts it, "the arrangement," or the
relations, of the ideas are different from those in
which the impressions, whence these ideas are derived, occurred ; or in other words, that the thing
imagined has not happened.   In popular usage,
however, imagination is frequently employed for
simple memory—"In imagination I was back in
the old times."

It is a curious omission on Hume's part that
while thus dwelling on two classes of ideas,
Memories and *Imaginations*, he has not, at the

same time, taken notice of a third group, of no small importance, which are as different from imaginations as memories are; though, like the latter, they are often confounded with pure imaginations in general speech. These are the ideas of expectation, or as they may be called for the sake of brevity, *Expectations;* which differ from simple imaginations in being associated with the idea of the existence of corresponding impressions, in the future, just as memories contain the idea of the existence of the corresponding impressions in the past.

The ideas belonging to two of the three groups enumerated: namely, memories and expectations, present some features of particular interest. And first, with respect to memories.

In Hume's words, all simple ideas are copies of simple impressions. The idea of a single sensation is a faint, but accurate, image of that sensation; the idea of a relation is a reproduction of the feeling of co-existence, of succession, or of similarity. But, when complex impressions or complex ideas are reproduced as memories, it is probable that the copies never give all the details of the originals with perfect accuracy, and it is certain that they rarely do so. No one possesses a memory so good, that if he has only once observed a natural object, a second inspection does not show him something that he has forgotten. Almost all, if not all, our memories are therefore

sketches, rather than portraits, of the originals—
the salient features are obvious, while the subordinate characters are obscure or unrepresented.

Now, when several complex impressions which
are more or less different from one another—let
us say that out of ten impressions in each, six are
the same in all, and four are different from all
the rest—are successively presented to the mind,
it is easy to see what must be the nature of the
result. The repetition of the six similar impressions will strengthen the six corresponding
elements of the complex idea, which will therefore acquire greater vividness; while the four
differing impressions of each will not only acquire
no greater strength than they had at first, but, in
accordance with the law of association, they will
all tend to appear at once, and will thus neutralise
one another.

This mental operation may be rendered com-
prehensible by considering what takes place in
the formation of compound photographs—when
the images of the faces of six sitters, for
example, are each received on the same photo-
graphic plate, for a sixth of the time requisite
to take one portrait. The final result is that all
those points in which the six faces agree are
brought out strongly, while all those in which they
differ are left vague; and thus what may be
termed a *generic* portrait of the six, in contradis-
tinction to a *specific* portrait of any one, is produced.

Thus our ideas of single complex impressions are incomplete in one way, and those of numerous, more or less similar, complex impressions are incomplete in another way; that is to say, they are *generic*, not *specific*. And hence it follows, that our ideas of the impressions in question are not, in the strict sense of the word, copies of those impressions; while at the same time, they may exist in the mind independently of language.

The generic ideas which are formed from several similar, but not identical, complex experiences are what are commonly called *abstract* or *general* ideas; and Berkeley endeavoured to prove that all general ideas are nothing but particular ideas annexed to a certain term, which gives them a more extensive signification, and makes them recall, upon occasion, other individuals which are similar to them. Hume says that he regards this as " one of the greatest and the most valuable discoveries that has been made of late years in the republic of letters," and endeavours to confirm it in such a manner that it shall be " put beyond all doubt and controversy."

I may venture to express a doubt whether he has succeeded in his object; but the subject is an abstruse one ; and I must content myself with the remark, that though Berkeley's view appears to be largely applicable to such general ideas as are formed after language has been acquired, and to all the more abstract sort of conceptions, yet that general ideas of sensible objects may nevertheless be produced in the way indicated, and may exist independently of language. In dreams, one sees houses, trees and other objects, which are perfectly recognisable as such, but which remind one of the actual objects as seen "out of the corner of the eye," or of the pictures thrown by a badly-focused magic lantern. A man addresses us who is like a figure seen by twilight; or we travel through countries where every feature of the scenery is vague; the outlines of the hills are ill-marked, and the rivers have no defined banks. They are, in short, generic ideas of many past impressions of men, hills, and rivers. An anatomist who occupies himself intently with the examination of several specimens of some new kind of animal, in course of time acquires so vivid a conception of its form and structure, that the idea may take visible shape and become a sort of waking dream. But the figure which thus presents itself is generic, not specific. It is no copy of any one specimen, but, more or less, a mean of the series; and there seems no reason to doubt that the minds of children before they learn to speak, and of deaf mutes, are people with similarly generated generic ideas of sensible objects.

It has been seen that a memory is a complex idea made up of at least two constituents. In the

first place there is the idea of an object; and
secondly, there is the idea of the relation of antecedents between that object and some present
objects.

To say that one has a recollection of a given
event and to express the belief that it happened,
are two ways of giving an account of one and the
same mental fact. But the former mode of stating the fact of memory is preferable, at present,
because it certainly does not presuppose the existence of language in the mind of the rememberer ;
while it may be said that the latter does. It is
perfectly possible to have the idea of an event A,
and of the events B, C, D, which came between it
and the present state E, as mere mental pictures.
It is hardly to be doubted that children have very
distinct memories long before they can speak ; and
we believe that such is the case because they act
upon their memories. But, if they act upon their
memories, they to all intents and purposes believe
their memories. In other words, though, being
devoid of language, the child cannot frame a proposition expressive of belief; cannot say " sugar-
plum was sweet "; yet the physical operation of
which that proposition is merely the verbal expression, is perfectly effected. The experience of
the co-existence of sweetness with sugar has produced a state of mind which bears the same relation
to a verbal proposition, as the natural disposition
to produce a given idea, assumed to exist by

Descartes as an "innate idea" would bear to that idea put into words.

The fact that the beliefs of memory precede the use of language, and therefore are originally purely instinctive, and independent of any rational justification, should have been of great importance to Hume, from its bearing upon his theory of causation; and it is curious that he has not adverted to it, but always takes the trustworthiness of memories for granted. It may be worth while briefly to make good the omission.

That I was in pain, yesterday, is as certain to me as any matter of fact can be; by no effort of the imagination is it possible for me really to entertain the contrary belief. At the same time, I am bound to admit, that the whole foundation for my belief is the fact, that the idea of pain is indissolubly associated in my mind with the idea of that much past time. Any one who will be at the trouble may provide himself with hundreds of examples to the same effect.

This and similar observations are important under another aspect. They prove that the idea of even a single strong impression may be so powerfully associated with that of a certain time, as to originate a belief of which the contrary is inconceivable, and which may therefore be properly said to be necessary. A single weak, or moderately strong, impression may not be represented by any memory. But this defect of weak experiences may be compensated by their repetition; and what Hume means by "custom" or "habit" is simply the repetition of experiences.

"Wherever the repetition of any particular act or operation produces a propensity to renew the same act or operation, without being impelled by any reasoning or process of the understanding, we always say that this propensity is the effect of *Custom*. By employing that word, we pretend not to have given the ultimate reason of such a propensity. We only point out a principle of human nature which is universally acknowledged, and which is well known by its effects."—(IV. p. 52.)

It has been shown that an expectation is a complex idea which, like a memory, is made up of two constituents. The one is the idea of an object, the other is the idea of a relation of sequence between that object and some present object; and the reasoning which applied to memories applies to expectations. To have an expectation [1] of a given event, and to believe that it will happen, are only two modes of stating the same fact. Again, just in the same way as we call a memory, put into words, a belief, so we give the same name to an expectation in like clothing. And the fact already cited, that a child before it can speak acts upon its memories, is good evidence that it forms expectations. The infant who knows the meaning neither of " sugar-plum " nor

[1] We give no name to faint memories; but expectations of like character play so large a part in human affairs, that they, together with the associated emotions of pleasure and pain, are distinguished as " hopes " or " fears."

of "sweet," nevertheless is in full possession of that complex idea, which, when he has learned to employ language, will take the form of the verbal proposition, " A sugar-plum will be sweet."

Thus, beliefs of expectation, or at any rate their potentialities, are, as much as those of memory, antecedent to speech, and are as incapable of justification by any logical process. In fact, expectations are but memories inverted. The association which is the foundation of expectation must exist as a memory before it can play its part. As Hume says,—

". . . it is certain we here advance a very intelligible pro-position at least, if not a true one, when we assert that after the constant conjunction of two objects, heat and flame, for instance, weight and solidity, we are determined by custom alone to ex-pect the one from the appearance of the other.   This hypothesis seems even the only one which explains the difficulty why we draw from a thousand instances, an inference which we are not able to draw from one instance, that is in no respect different from them." . . .

"Custom, then, is the great guide of human life.   It is that principle alone which renders our experience useful to us, and makes us expect, for the future, a similar train of events with those which have appeared in the past." . . .

"All belief of matter-of-fact or real existence is derived merely from some object present to the memory or senses, and a customary conjunction between that and some other object ; or in other words, having found, in many instances, that any two kinds of objects, flame and heat, snow and cold, have always been conjoined together, if flame or snow be presented anew to the senses, the mind is carried by custom to expect heat or cold, and to *believe* that such a quality does exist and will discover itself upon a nearer approach.   This belief is the necessary result of placing the mind in such circumstances. It is an operation of the soul, when we are so situated, as unavoidable as to feel the passion of love, when we receive benefits, or hatred, when we meet with injuries. All these operations are a species of natural instincts, which no reasoning or process of the thought and understanding is able either to produce or to prevent."— (IV. pp. 52—56.)

The only comment that appears needful here is, that Hume has attached somewhat too exclusive a weight to that repetition of experiences to which alone the term "custom" can be properly applied. The proverb says that "a burnt child dreads the fire"; and any one who will make the experiment will find, that one burning is quite sufficient to establish an indissoluble belief that contact with fire and pain go together.

As a sort of inverted memory, expectation follows the same laws; hence, while a belief of expectation is, in most cases, as Hume truly says, established by custom, or the repetition of weak impressions, it may quite well be based upon a single strong experience. In the absence of language, a specific memory cannot be strengthened by repetition. It is obvious that that which has happened cannot happen again, with the same collateral associations of co-existence and succession. But, memories of the co-existence and succession of impresions are capable of being indefinitely strengthened by the recurrence of similar impressions, in the same order, even though the collateral associations are totally

different ; in fact, the ideas of these impressions
become generic.

If I recollect that a piece of ice was cold yesterday, nothing can strengthen the recollection of
that particular fact ; on the contrary, it may grow
weaker, in the absence of any record of it. But
if I touch ice to-day and again find it cold, the
association is repeated, and the memory of it
becomes stronger. And by this very simple
process of repetition of experience, it has become
utterly impossible for us to think of having
handled ice without thinking of its coldness. But,
that which is, under the one aspect, the strengthening of a memory, is, under the other, the inten-
sification of an expectation. Not only can we not
think of having touched ice, without feeling cold,
but we cannot think of touching ice, in the future,
without expecting to feel cold. An expectation so
strong that it cannot be changed, or abolished,
may thus be generated out of repeated experiences.
And it is important to note that such expectations may be formed quite unconsciously. In my
dressing-room, a certain can is usually kept full of
water, and I am in the habit of lifting it to pour
out water for washing. Sometimes the servant
has forgotten to fill it, and then I find that, when
I take hold of the handle, the can goes up with a
jerk. Long association has, in fact, led me to
expect the can to have a considerable weight ; and,

quite unawares, my muscular effort is adjusted to
the expectation.

The process of strengthening generic memories
of succession, and, at the same time, intensifying
expectations of succession, is what is commonly
called *verification*. The impression B has frequently been observed to follow the impression A.
The association thus produced is represented as
the memory, A -> B. When the impression A
appears again, the idea of B follows, associated
with that of the immediate appearance of
the impression B. If the impression B does
appear, the expectation is said to be verified;
while the memory A -> B is strengthened, and
gives rise in turn to a stronger expectation. And
repeated verification may render that expectation
so strong that its non-verification is inconceivable.

CHAPTER V

In the course of the preceding chapters, attention has been more than once called to the fact, that the elements of consciousness and the operations of the mental faculties, under discussion, exist independently of and antecedent to, the existence of language.

If any weight is to be attached to arguments from analogy, there is overwhelming evidence in favour of the belief that children, before they can speak, and deaf mutes, possess the feelings to which those who have acquired the faculty of speech apply the name of sensations; that they have the feelings of relation ; that trains of ideas pass through their minds ; that generic ideas are formed from specific ones; and, that among these, ideas of memory and expectation occupy a most important place, inasmuch as, in their quality of potential beliefs, they furnish the grounds of action. This conclusion, in truth, is one of those which, though they cannot be demonstrated, are never doubted; and, since it is highly probable and cannot be disproved, we are quite safe in accepting it, as, at any rate, a good working hypothesis.

But, if we accept it, we must extend it to a much wider assemblage of living beings. Whatever cogency is attached to the arguments in favour of the occurrence of all the fundamental phenomena of mind in young children and deaf mutes, an equal force must be allowed to appertain to those which may be adduced to prove that the higher animals have minds. We must admit that Hume does not express himself too strongly when he says—

"no truth appears to me more evident, than that the beasts are endowed with thought and reason as well as men. The arguments are in this case so obvious, that they never escape the most stupid and ignorant."—(I. p. 232.)

In fact, this is one of the few cases in which the conviction which forces itself upon the stupid and the ignorant, is fortified by the reasonings of the intelligent, and has its foundation deepened by every increase of knowledge. It is not merely that the observation of the actions of animals almost irresistibly suggests the attribution to them of mental states, such as those which accompany corresponding actions in men. The minute comparison which has been instituted by anatomists and physiologists between the organs which we know to constitute the apparatus of thought in man, and the corresponding organs in brutes, has

demonstrated the existence of the closest similarity between the two, not only in structure, as
far as the microscope will carry us, but in function, as far as functions are determinable by
experiment. There is no question in the mind of
any one acquainted with the facts that, so far as
observation and experiment can take us, the
structure and the functions of the nervous system
are fundamentally the same in an ape, or in a dog,
and in a man. And the suggestion that we must
stop at the exact point at which direct proof fails
us; and refuse to believe that the similarity which
extends so far stretches yet further, is no better
than a quibble. Robinson Crusoe did not feel
bound to conclude, from the single human foot-
print which he saw in the sand, that the maker of
the impression had only one leg.

Structure for structure, down to the minutest
microscopical details, the eye, the ear, the
olfactory organs, the nerves, the spinal cord, the
brain of an ape, or of a dog, correspond with the
same organs in the human subject. Cut a nerve,
and the evidence of paralysis, or of insensibility,
is the same in the two cases; apply pressure to
the brain, or administer a narcotic, and the signs
of intelligence disappear in the one as in the other.
Whatever reason we have for believing that the
changes which take place in the normal cerebral
substance of man give rise to states of consciousness, the same reason exists for the belief that

the modes of motion of the cerebral substance of an ape, or of a dog, produce like effects.

A dog acts as if he had all the different kinds of impressions of sensation of which each of us is cognisant. Moreover, he governs his movements exactly as if he had the feelings of distance, form, succession, likeness, and unlikeness, with which we are familiar, or as if the impressions of relation were generated in his mind as they are in our own. Sleeping dogs frequently appear to dream. If they do, it must be admitted that ideation goes on in them while they are asleep; and, in that case, there is no reason to doubt that they are conscious of trains of ideas in their waking state. Further, that dogs, if they possess ideas at all, have memories and expectations, and those potential beliefs of which these states are the foundation, can hardly be doubted by any one who is conversant with their ways. Finally, there would appear to be no valid argument against the supposition that dogs form generic ideas of sensible objects. One of the most curious peculiarities of the dog mind is its inherent snobbishness, shown by the regard paid to external respectability. The dog who barks furiously at a beggar will let a well-dressed man pass him without opposition. Has he not then a " generic idea" of rags and dirt associated with the idea of aversion, and that of sleek broadcloth associated with the idea of liking?

In short, it seems hard to assign any good reason for denying to the higher animals any mental state, or process, in which the employment of the vocal or visual symbols of which language is composed is not involved ; and comparative psychology confirms the position in relation to the rest of the animal world assigned to man by comparative anatomy. As comparative anatomy is easily able to show that, physically, man is but the last term of a long series of forms, which lead, by slow gradations, from the highest mammal to the almost formless speck of living protoplasm, which lies on the shadowy boundary between animal and vegetable life ; so, comparative psychology, though but a young science, and far short of her elder sister's growth, points to the same conclusion.

In the absence of a distinct nervous system, we have no right to look for its product, consciousness : and, even in those forms of animal life in which the nervous apparatus has reached no higher degree of development, than that exhibited by the system of the spinal cord and the foundation of the brain in ourselves, the argument from analogy leaves the assumption of the existence of any form of consciousness unsupported. With the super-addition of a nervous apparatus corresponding with the cerebrum in ourselves, it is allowable to suppose the appearance of the simplest states of consciousness, or the sensations ;

and it is conceivable that these may at first exist, without any power of reproducing them, as memories; and, consequently, without ideation. Still higher, an apparatus of correlation may be superadded, until, as all these organs become more developed, the condition of the highest speechless animals is attained.

It is a remarkable example of Hume's sagacity that he perceived the importance of a branch of science which, even now, can hardly be said to exist; and that, in a remarkable passage, he sketches in bold outlines the chief features of comparative psychology.

". . . any theory, by which we explain the operations of the understanding, or the origin and connection of the passions in man, will acquire additional authority if we find that the same theory is requisite to explain the same phenomena in all other animals. We shall make trial of this with regard to the hypothesis by which we have, in the foregoing discourse, endeavoured to account for all experimental reasonings ; and it is hoped that this new point of view will serve to confirm all our former observations.

" *First*, it seems evident that animals, as well as men, learn many things from experience, and infer that the same events will always follow from the same causes. By this principle they become acquainted with the more obvious properties of external objects, and gradually, from their birth, treasure up a knowledge of the nature of fire, water, earth, stones, heights, depths, &c., and of the effects which result from their operation. The ignorance and inexperience of the young are here plainly distinguishable from the cunning and sagacity of the old, who have learned, by long observation, to avoid what hurt them, and pursue what gave ease or pleasure. A horse that has been accustomed to the field, becomes acquainted with the proper

height which he can leap, and will never attempt what exceeds
his force and ability.   An old greyhound will trust the more
fatiguing part of the chase to the younger, and will place himself so as to meet the hare in her doubles ; nor are the conjectures
which he forms on this occasion founded on anything but his
observation and experience.

" This is still more evident from the effects of discipline and
education on animals, who, by the proper application of rewards
and punishments, may be taught any course of action, the most
contrary to their natural instincts and propensities.   Is it not
experience which renders a dog apprehensive of pain when you
menace him, or lift up the whip to beat him ?   Is it not even
experience which makes him answer to his name, and infer from
such an arbitrary sound that you mean him rather than any of
his fellows, and intend to call him, when you pronounce it in a
certain manner and with a certain tone and accent ?

" In all these cases we may observe that the animal infers
some fact beyond what immediately strikes his senses ; and that
this inference is altogether founded on past experience, while the
creature expects from the present object the same consequences
which it has always found in its observation to result from
similar objects.

" Secondly, it is impossible that this inference of the animal
can be founded on any process of argument or reasoning by
which he concludes that like events must follow like objects,
and that the course of nature will always be regular in its
operations.   For if there be in reality any arguments of this
nature they surely lie too abstruse for the observation of such
imperfect understandings ; since it may well employ the utmost
care and attention of a philosophic genius to discover and observe
them.   Animals therefore are not guided in these inferences by
reasoning ; neither are children ; neither are the generality of
mankind in their ordinary actions and conclusions ; neither are
philosophers themselves, who, in all the active parts of life, are
in the main the same as the vulgar, and are governed by the
same maxims.   Nature must have provided some other principle,
of more ready and more general use and application ; nor can an
operation of such immense consequence in life as that of in-

ferring effects from causes, be trusted to the uncertain process of
reasoning and argumentation. Were this doubtful with regard
to men, it seems to admit of no question with regard to the
brute creation ; and the conclusion being once firmly established
in the one, we have a strong presumption, from all the rules of
analogy, that it ought to be universally admitted, without any
exception or reserve. It is custom alone which engages animals,
from every object that strikes their senses, to infer its usual
attendant, and carries their imagination from the appearance of
the one to conceive the other, in that particular manner which
we denominate *belief*. No other explication can be given of
this operation in all the higher as well as lower classes of sensitive beings which fall under our notice and observation."
—(IV. pp. 122—4.)

It will be observed that Hume appears to
contrast the "inference of the animal" with the
" process of argument or reasoning in man." But
it would be a complete misapprehension of his
intention, if we were to suppose, that he thereby
means to imply that there is any real difference
between the two processes. The "inference of
the animal" is a potential belief of expectation;
the process of argument, or reasoning in man is
based upon potential beliefs of expectation, which
are formed in the man exactly in the same way as
in the animal. But, in men endowed with speech
the mental state which constitutes the potential
belief is represented by a verbal proposition, and
thus becomes what all the world recognises as a
belief. The fallacy which Hume combats is, that
the proposition, or verbal representative of a
belief, has come to be regarded as a reality,

instead of as the mere symbol which it really is; and that reasoning, or logic, which deals with nothing but propositions, is supposed to be necessary in order to validate the natural fact symbolised by those propositions.. It is a fallacy similar to that of supposing that money is the foundation of wealth, whereas it is only the wholly unessential symbol of property.

In the passage which immediately follows that just quoted, Hume makes admissions which might be turned to serious account against some of his own doctrines.

" But though animals learn many parts of their knowledge from observation, there are also many parts of it which they derive from the original hand of Nature, which much exceed the share of capacity they possess on ordinary occasions, and in which they improve, little or nothing, by the longest practice and experience. These we denominate INSTINCTS, and are so apt to admire as something very extraordinary and inexplicable by all the disquisitions of human understanding. But our wonder will perhaps cease or diminish when we consider that the experimental reasoning itself, which we possess in common with beasts, and on which the whole conduct of life depends, is nothing but a species of instinct or mechanical power, that acts in us unknown to ourselves, and in its chief operations is not directed by any such relations or comparison of ideas as are the proper objects of our intellectual faculties.

" Though the instinct be different, yet still it is an instinct which teaches a man to avoid the fire, as much as that which teaches a bird, with such exactness, the art of incubation and the whole economy and order of its nursery."—(IV. pp. 125, 126.)

The parallel here drawn between the " avoidance of a fire " by a man and the incubatory instinct of a bird is inexact. The man avoids fire when he has had experience of the pain produced by burning; but the bird incubates the first time it lays eggs, and therefore before it has had any experience of incubation. For the comparison to be admissible, it would be necessary that a man should avoid fire the first time he saw it, which is notoriously not the case.

The term "instinct" is very vague and ill-defined. It is commonly employed to denote any action, or even feeling, which is not dictated by conscious reasoning, whether it is, or is not, the result of previous experience. It is "instinct" which leads a chicken just hatched to pick up a grain of corn; parental love is said to be "instinctive"; the drowning man who catches at a straw does it "instinctively"; and the hand that accidentally touches something hot is drawn back by "instinct." Thus "instinct" is made to cover everything from a simple reflex movement, in which the organ of consciousness need not be at all implicated, up to a complex combination of acts directed towards a definite end and accompanied by intense consciousness.

But this loose employment of the term "instinct" really accords with the nature of the thing; for it is wholly impossible to draw any line of demarcation between reflex actions and instincts. If a frog, on the flank of which a little drop of acid has been placed, rubs it off with the foot of the same side; and, if that foot be held, performs the same operation, at the cost of much effort, with the other foot, it certainly displays a curious instinct. But it is no less true that the whole operation is a reflex operation of the spinal cord, which can be performed quite as well when the brain is destroyed; and between which and simple reflex actions there is a complete series of gradations. In like manner, when an infant takes the breast, it is impossible to say whether the action should be rather termed instinctive or reflex.

What are usually called the instincts of animals are, however, acts of such a nature that, if they were performed by men, they would involve the generation of a series of ideas and of inferences from them; and it is a curious, apparently an insoluble, problem whether they are, or are not, accompanied by cerebral changes of the same nature as those which give rise to ideas and inferences in ourselves. When a chicken picks up a grain, for example, are there, firstly, certain sensations, accompanied by the feeling of relation between the grain and its own body; secondly, a desire of the grain; thirdly, a volition to seize it? Or, are only the sensational terms of the series actually represented in consciousness?

The latter seems the more probable opinion, though it must be admitted that the other alternative is possible. But, in this case, the series of mental states which occurs is such as would be represented in language by a series of propositions, and would afford proof positive of the existence of innate ideas, in the Cartesian sense. Indeed, a metaphysical fowl, brooding over the mental operations of his fully-fledged consciousness, might appeal to the fact as proof that, in the very first action of his life, he assumed the existence of the Ego and the non-Ego, and of a relation between the two.

In all seriousness, if the existence of instincts be granted, the possibility of the existence of innate ideas, in the most extended sense ever imagined by Descartes, must also be admitted. In fact, Descartes, as we have seen, illustrates what he means by an innate idea, by the analogy of hereditary diseases or hereditary mental peculiarities, such as generosity. On the other hand, hereditary mental tendencies may justly be termed instincts; and still more appropriately might those special proclivities, which constitute what we call genius, come into the same category.

The child who is impelled to draw as soon as it can hold a pencil; the Mozart who breaks out into music as early; the boy Bidder who worked out the most complicated sums without learning arithmetic; the boy Pascal who evolved Euclid out of his own consciousness: all these may be said to have been impelled by instinct, as much as are the beaver and the bee. And the man of

genius is distinct in kind from the man of clever-
ness, by reason of the working within him of
strong innate tendencies—which cultivation may
improve, but which it can no more create, than
horticulture can make thistles bear figs.   The
analogy between a musical instrument and the
mind holds good here also.   Art and industry may
get much music, of a sort, out of a penny whistle;
but, when all is done, it has no chance against an
organ.   The innate musical potentialities of the
two are infinitely different.

CHAPTER VI

THOUGH we may accept Hume's conclusion that speechless animals think, believe, and reason; yet, it must be borne in mind, that there is an important difference between the signification of the terms when applied to them and when applied to those animals which possess language. The thoughts of the former are trains of mere feelings; those of the latter are, in addition, trains of the ideas of the signs which represent feelings, and which are called " words."

A word, in fact, is a spoken or written sign, the idea of which is, by repetition, so closely associated with the idea of the simple or complex feeling which it represents, that the association becomes indissoluble. No Englishman, for example, can think of the word " dog " without immediately having the idea of the group of impressions to which that name is given; and conversely, the group of impressions immediately calls up the idea of the word "dog."

The association of words with impressions and ideas is the process of naming; and language approaches perfection, in proportion as the shades of difference between various ideas and impressions are represented by differences in their names.

The names of simple impressions and ideas, or of groups of co-existent or successive complex impressions and ideas, considered *per se*, are substantives; as redness, dog, silver, mouth; while the names of impressions or ideas considered as parts or attributes of a complex whole, are adjectives. Thus redness, considered as part of the complex idea of a rose, becomes the adjective red; flesh-eater, as part of the idea of a dog, is represented by carnivorous; whiteness, as part of the idea of silver, is white; and so on.

The linguistic machinery for the expression of belief is called *predication;* and, as all beliefs express ideas of relation, we may say that the sign of predication is the verbal symbol of a feeling of relation. The words which serve to indicate predication are verbs. If I say "silver" and then "white," I merely utter two names; but if I interpose between them the verb "is," I express a belief in the co-existence of the feeling of whiteness with the other feelings which constitute the totality of the complex idea of silver; in other words, I predicate "whiteness" of silver.

In such a case as this, the verb expresses predication and nothing else, and is called a copula.
But, in the great majority of verbs, the word is
the sign of a complex idea, and the predication is
expressed only by its form. Thus in "silver
shines," the verb "to shine" is the sign for the
feeling of brightness, and the mark of predication
lies in the form "shine-*s.*"

Another result is brought about by the forms
of verbs. By slight modifications they are made
to indicate that a belief, or predication, is a
memory, or is an expectation. Thus "silver
shone" expresses a memory; "silver *will* shine"
an expectation.

The form of words which expresses a predication
is a proposition. Hence, every predication is the
verbal equivalent of a belief; and, as every belief is
either an immediate consciousness, a memory, or an
expectation, and as every expectation is traceable
to a memory, it follows that, in the long run, all
propositions express either immediate states of
consciousness, or memories. The proposition
which predicates A of X must mean either, that
the fact is testified by my present consciousness,
as when I say that two colours, visible at this
moment, resemble one another; or that A is
indissolubly associated with X in memory; or that
A is indissolubly associated with X in expectation.
But it has already been shown that expectation
is only an expression of memory.

Hume does not discuss the nature of language, but so much of what remains to be said, concerning his philosophical tenets, turns upon the value and the origin of verbal propositions, that this summary sketch of the relations of language to the thinking process will probably not be deemed superfluous.

So large an extent of the field of thought is traversed by Hume, in his discussion of the verbal propositions in which mankind enshrine their beliefs, that it would be impossible to follow him throughout all the windings of his long journey, within the limits of this essay. I purpose, therefore, to limit myself to those propositions which concern—1. Necessary Truths; 2. The order of Nature ; 3. The Soul; 4. Theism; 5. The Passions and Volition; 6. The Principle of Morals.

Hume's views respecting necessary truths, and more particularly concerning causation, have, more than any other part of his teaching, contributed to give him a prominent place in the history of philosophy.

" All the objects of human reason and inquiry may naturally be divided into two kinds, to wit, *relations of ideas* and *matters of fact.* Of the first kind are the sciences of geometry, algebra, and arithmetic, and, in short, every affirmation which is either intuitively or demonstratively certain. *That the square of the hypothenuse is equal to the square of the two sides*, is a proposition which expresses a relation between these two figures. *That three times five is equal to the half of thirty*, expresses a relation between these numbers. Propositions of this kind are discoverable by the mere operation of thought without dependence on whatever is anywhere existent in the universe. Though there never were a circle or a triangle in nature, the truths demonstrated by Euclid would for ever retain their certainty and evidence.

"Matters of fact, which are the second objects of human reason, are not ascertained in the same manner, nor is an evidence of their truth, however great, of a like nature with the foregoing. The contrary of every matter of fact is still possible, because it can never imply a contradiction, and is conceived by the mind with the same facility and distinctness, as if ever so conformable to reality. *That the sun will not rise to-morrow*, is no less intelligible a proposition, and implies no more contradiction, than the affirmation, *that it will rise*. We should in vain, therefore, attempt to demonstrate its falsehood. Were it demonstratively false, it would imply a contradiction, and could never be distinctly conceived by the mind."—(IV. pp. 32, 33.)

The distinction here drawn between the truths of geometry and other kinds of truth is far less sharply indicated in the "Treatise," but as Hume expressly disowns any opinions on these matters but such as are expressed in the "Inquiry," we may confine ourselves to the latter; and it is needful to look narrowly into the propositions here laid down, as much stress has been laid upon Hume's admission that the truths of mathematics are intuitively and demonstratively certain; in other words, that they are necessary and, in that respect, differ from all other kinds of belief.

What is meant by the assertion that "propositions of this kind are discoverable by the mere operation of thought without dependence on what is anywhere existent in the universe "?

Suppose that there were no such things as impressions of sight and touch anywhere in the universe, what idea could we have even of a straight line, much less of a triangle and of the relations between its sides? The fundamental proposition of all Hume's philosophy is that ideas are copied from impressions; and, therefore, if there were no impressions of straight lines and triangles there could be no ideas of straight lines and triangles. But what we mean by the universe is the sum of our actual and possible impressions.

So, again, whether our conception of number is derived from relations of impressions in space or in time, the impressions must exist in nature, that is, in experience, before their relations can be perceived. Form and number are mere names for certain relations between matters of fact; unless a man had seen or felt the difference between a straight line and a crooked one, straight and crooked would have no more meaning to him, than red and blue to the blind.

The axiom, that things which are equal to the same are equal to one another, is only a particular case of the predication of similarity; if there were no impressions, it is obvious that there could be no predicates. But what is an existence in the universe but an impression?

If what are called necessary truths are rigidly analysed, they will be found to be of two kinds. Either they depend on the convention which underlies the possibility of intelligible speech, that terms shall always have the same meaning; or they are propositions the negation of which implies the dissolution of some association in memory or expectation, which is in fact indissoluble; or the denial of some fact of immediate consciousness.

The "necessary truth" A = A means that the perception which is called A shall always be called A. The "necessary truth" that "two straight lines cannot inclose a space," means that we have no memory, and can form no expectation of their so doing. The denial of the "necessary truth" that the thought now in my mind exists, involves the denial of consciousness.

To the assertion that the evidence of matter of fact is not so strong as that of relations of ideas, it may be justly replied, that a great number of matters of fact are nothing but relations of ideas. If I say that red is unlike blue, I make an assertion concerning a relation of ideas; but it is also matter of fact, and the contrary proposition is inconceivable. If I remember [1] something that happened five minutes ago, that is matter of fact; and, at the same time, it expresses a relation

[1] Hume, however, expressly includes the "records of our memory" among his matters of fact.—(IV. p. 33.)

between the event remembered and the present time. It is wholly inconceivable to me that the event did not happen, so that my assurance respecting it is as strong as that which I have respecting any other necessary truth. In fact, the man is either very wise, or very virtuous, or very lucky, perhaps all three, who has gone through life without accumulating a store of such necessary beliefs which he would give a good deal to be able to disbelieve.

It would be beside the mark to discuss the matter further on the present occasion. It is sufficient to point out that, whatever may be the differences between mathematical and other truths, they do not justify Hume's statement. And it is, at any rate, impossible to prove that the cogency of mathematical first principles is due to anything more than these circumstances; that the experiences with which they are concerned are among the first which arise in the mind; that they are so incessantly repeated as to justify us, according to the ordinary laws of ideation, in expecting that the associations which they form will be of extreme tenacity; while the fact, that the expectations based upon them are always verified, finishes the process of welding them together.

Thus, if the axioms of mathematics are innate, nature would seem to have taken unnecessary trouble; since the ordinary process of association appears to be amply sufficient to confer upon them all the universality and necessity which they actually possess.

Whatever needless admissions Hume may have made respecting other necessary truths he is quite clear about the axiom of causation, "That whatever event has a beginning must have a cause;" whether and in what sense it is a necessary truth; and, that question being decided, whence it is derived.

With respect to the first question, Hume denies that it is a necessary truth, in the sense that we are unable to conceive the contrary. The evidence by which he supports this conclusion in the "Inquiry," however, is not strictly relevant to the issue.

"No object ever discovers, by the qualities which appear to the senses, either the cause which produced it, or the effects which will arise from it; nor can our reason, unassisted by experience, ever draw any inference concerning real existence and matter of fact."—(IV. p. 35.)

Abundant illustrations are given of this assertion, which indeed cannot be seriously doubted; but it does not follow that, because we are totally unable to say what cause preceded, or what effect will succeed, any event, we do not necessarily suppose that the event had a cause and will be succeeded by an effect. The scientific investigator who notes a new phenomenon may be utterly ignorant of its cause, but he will, without hesitation, seek for that cause. If you ask him why he does so, he will probably say that it must have had a cause; and thereby imply that his belief in causation is a necessary belief.

In the " Treatise " Hume indeed takes the bull by the horns:

". . . as all distinct ideas are separable from each other, and as the ideas of cause and effect are evidently distinct, 'twill be easy for us to conceive any object to be non-existent this moment and existent the next, without conjoining to it the distinct idea of a cause or productive principle."—(I. p. 111.)

If Hume had been content to state what he believed to be matter of fact, and had abstained from giving superfluous reasons for that which is susceptible of being proved or disproved only by personal experience, his position would have been stronger. For it seems clear that, on the ground of observation, he is quite right. Any man who lets his fancy run riot in a waking dream, may experience the existence at one moment, and the non-existence at the next, of phenomena which suggest no connexion of cause and effect. Not only so, but it is notorious that, to the unthinking mass of mankind, nine-tenths of the facts of life do not suggest the relation of cause and effect; and they practically deny the existence of any such relation by attributing them to chance. Few gamblers but would stare if they were told that the falling of a die on a particular face is as

much the effect of a definite cause as the fact of its
falling; it is a proverb that "the wind bloweth
where it listeth"; and even thoughtful men usually
receive with surprise the suggestion, that the
form of the crest of every wave that breaks, wind-
driven, on the sea-shore, and the direction of
every particle of foam that flies before the gale,
are the exact effects of definite causes; and, as
such, must be capable of being determined, de-
ductively, from the laws of motion and the properties of air and water.   So again, there are
large numbers of highly intelligent persons who
rather pride themselves on their fixed belief that
our volitions have no cause; or that the will
causes itself, which is either the same thing, or a
contradiction in terms.

Hume's argument in support of what appears
to be a true proposition, however, is of the circular
sort, for the major premiss, that all distinct ideas
are separable in thought, assumes the question at
issue.

But the question whether the idea of causation
is necessary, or not, is really of very little importance.   For, to say that an idea is necessary is
simply to affirm that we cannot conceive the contrary; and the fact that we cannot conceive the
contrary of any belief may be a presumption, but
is certainly no proof, of its truth.

In the well-known experiment of touching a
single round object, such as a marble, with crossed

fingers, it is utterly impossible to conceive that
we have not two round objects under them ; and,
though light is undoubtedly a mere sensation
arising in the brain, it is utterly impossible to
conceive that it is not outside the retina. In
the same way, he who touches anything with a
rod, not only is irresistibly led to believe that the
sensation of contact is at the end of the rod, but
is utterly incapable of conceiving that this sensation is really in his head. Yet that which is
inconceivable is manifestly true in all these cases.
The beliefs and the unbeliefs are alike necessary,
and alike erroneous.

It is commonly urged that the axiom of causation
cannot be derived from experience, because experience only proves that many things have causes,
whereas the axiom declares that all things have
causes. The syllogism, " many things which come
into existence have causes. A has come into
existence : therefore A had a cause," is obviously
fallacious, if A is not previously shown to be one
of the " many things." And this objection is
perfectly sound so far as it goes. The axiom of
causation cannot possibly be deduced from any
general proposition which simply embodies experience. But it does not follow that the belief,
or expectation, expressed by the axiom, is not a
product of experience, generated antecedently to,
and altogether independently of, the logically un-
justifiable language in which we express it.

In fact, the axiom of causation resembles all other beliefs of expectation in being the verbal symbol of a purely automatic act of the mind, which is altogether extra-logical, and would be illogical, if it were not constantly verified by experience. Experience, as we have seen, stores up memories; memories generate expectations or beliefs—why they do so may be explained hereafter by proper investigation of cerebral physiology. But to seek for the reason of the facts in the verbal symbols by which they are expressed, and to be astonished that it is not to be found there, is surely singular; and what Hume did was to turn attention from the verbal proposition to the psychical fact of which it is the symbol.

"When any natural object or event is presented, it is impossible for us, by any sagacity or penetration, to discover, or even conjecture, without experience, what event will result from it, or to carry our foresight beyond that object, which is immediately present to the memory and senses. Even after one instance or experiment, where we have observed a particular event to follow upon another, we are not entitled to form a general rule, or foretell what will happen in like cases; it being justly esteemed an unpardonable temerity to judge of the whole course of nature from one single experiment, however accurate or certain. But when one particular species of events has always, in all instances, been conjoined with another, we make no longer any scruple of foretelling one upon the appearance of the other, and of employing that reasoning which can alone assure us of any matter of fact or existence. We then call the one object *Cause*, the other *Effect*. We suppose that there is some connexion between them: some power in the one, by which it infallibly produces the other, and operates with the greatest certainty and strongest necessity. . . . But there is nothing in a number of instances, different from every single instance, which is supposed to be exactly similar ; except only, that after a repetition of similar instances, the mind is carried by habit, upon the appearance of one event, to expect its usual attendant, and to believe that it will exist. . . . The first time a man saw the communication of motion by impulse, as by the shock of two billiard balls, he could not pronounce that the one event was *connected*, but only that it was *conjoined*, with the other. After he has observed several instances of this nature, he then pronounces them to be *connected*. What alteration has happened to give rise to this new idea of *connexion ?* Nothing but that he now *feels* these events to be *connected* in his imagination, and can readily foresee the existence of the one from the appearance of the other. When we say, therefore, that one object is connected with another we mean only that they have acquired a connexion in our thought, and give rise to this inference, by which they become proofs of each other's exist-ence ; a conclusion which is somewhat extraordinary, but which seems founded on sufficient evidence."—(IV. pp. 87—89.)

In the fifteenth section of the third part of the " Treatise," under the head of the *Rules by which to Judge of Causes and Effects*, Hume gives a sketch of the method of allocating effects to their causes, upon which, so far as I am aware, no improvement was made down to the time of the publication of Mill's " Logic." Of Mill's four methods, that of *agreement* is indicated in the following passage :—

". . . where several different objects produce the same effect, it must be by means of some quality which we discover to be common amongst them. For as like effects imply like causes, we must always ascribe the causation to the circumstance wherein we discover the resemblance."—(I. p. 229.)

L 2

Next, the foundation of the *method of difference* is stated :—

" The difference in the effects of two resembling objects must proceed from that particular in which they differ. For, as like causes always produce like effects, when in any instance we find our expectation to be disappointed, we must conclude that this irregularity proceeds from some difference in the causes."— (I. p. 230.)

In the succeeding paragraph the *method of concomitant variations* is foreshadowed.

" When any object increases or diminishes with the increase or diminution of the cause, 'tis to be regarded as a compounded effect, derived from the union of the several different effects which arise from the several different parts of the cause. The absence or presence of one part of the cause is here supposed to be always attended with the absence or presence of a proportionable part of the effect. This constant conjunction sufficiently proves that the one part is the cause of the other. We must, however, beware not to draw such a conclusion from a few experiments."—(I. p. 230.)

Lastly, the following rule, though awkwardly stated, contains a suggestion of the *method of residues :*—

". . . an object which exists for any time in its full perfection without any effect, is not the sole cause of that effect, but requires to be assisted by some other principle, which may forward its influence and operation. For as like effects necessarily follow from like causes, and in a contiguous time and place, their separation for a moment shows that these causes are not complete ones."—(I. p. 230.)

In addition to the bare notion of necessary connexion between the cause and its effect, we un-

doubtedly find in our minds the idea of something
resident in the cause which, as we say, produces
the effect, and we call this something Force, Power,
or Energy. Hume explains Force and Power as
the results of the association with inanimate causes
of the feelings of endeavour or resistance which we
experience, when our bodies give rise to, or resist,
motion.

If I throw a ball, I have a sense of effort which
ends when the ball leaves my hand ; and, if I catch
a ball, I have a sense of resistance which comes
to an end with the quiescence of the ball. In the
former case, there is a strong suggestion of something having gone from myself into the ball ; in
the latter, of something having been received from
the ball. Let any one hold a piece of iron near a
strong magnet, and the feeling that the magnet
endeavours to pull the iron one way, in the same
manner as he endeavours to pull it in the opposite
direction, is very strong.

As Hume says :—

"No animal can put external bodies in motion without the
sentiment of a *nisus*, or endeavour ; and every animal has a
sentiment or feeling from the stroke or blow of an external
object that is in motion. These sensations, which are merely
animal, and from which we can, *à priori*, draw no inference, we
are apt to transfer to inanimate objects, and to suppose that they
have some such feelings whenever they transfer or receive
motion."—(IV. p. 91, *note.*)

It is obviously, however, an absurdity not less

gross than that of supposing the sensation of
warmth to exist in a fire, to imagine that the subjective sensation of effort, or resistance, in ourselves
can be present in external objects, when they stand
in the relation of causes to other objects.

To the argument, that we have a right to suppose the relation of cause and effect to contain
something more than invariable succession, because,
when we ourselves act as causes, or in volition, we
are conscious of exerting power; Hume replies,
that we know nothing of the feeling we call power
except as effort or resistance; and that we have
not the slightest means of knowing whether it has
anything to do with the production of bodily
motion or mental changes. And he points out,
as Descartes and Spinoza had done before him,
that when voluntary motion takes place, that
which we will is not the immediate consequence
of the act of volition, but something which is
separated from it by a long chain of causes and
effects. If the will is the cause of the movement
of a limb, it can be so only in the sense that the
guard who gives the order to go on, is the cause
of the transport of a train from one station to
another.

"We learn from anatomy, that the immediate object of power
in voluntary notion is not the member itself which is moved,
but certain muscles and nerves and animal spirits, and perhaps
something still more minute and unknown, through which the
motion is successively propagated, ere it reached the member

itself, whose motion is the immediate object of volition. Can there be a more certain proof that the power by which the whole operation is performed, so far from being directly and fully known by an inward sentiment or consciousness, is to the last degree mysterious and unintelligible ? Here the mind wills a certain event : Immediately another event, unknown to ourselves, and totally different from the one intended, is produced : This event produces another equally unknown : Till at last, through a long succession, the desired event is produced."—(IV. p. 78.)

A still stronger argument against ascribing an objective existence to force or power, on the strength of our supposed direct intuition of power in voluntary acts, may be urged from the unquestionable fact, that we do not know, and cannot know, that volition does cause corporeal motion; while there is a great deal to be said in favour of the view that it is no cause, but merely a concomitant of that motion. But the nature of volition will be more fitly considered hereafter

CHAPTER VII

IF our beliefs of expectation are based on our beliefs of memory, and anticipation is only inverted recollection, it necessarily follows that every belief of expectation implies the belief that the future will have a certain resemblance to the past. From the first hour of experience, onwards, this belief is constantly being verified, until old age is inclined to suspect that experience has nothing new to offer. And when the experience of generation after generation is recorded, and a single book tells us more than Methuselah could have learned, had he spent every waking hour of his thousand years in learning; when apparent disorders are found to be only the recurrent pulses of a slow working order, and the wonder of a year becomes the commonplace of a century; when repeated and minute examination never reveals a break in the chain of causes and effects; and the

whole edifice of practical life is built upon our
faith in its continuity; the belief, that that chain
has never been broken and will never be broken,
becomes one of the strongest and most justifiable
of human convictions. And it must be admitted
to be a reasonable request, if we ask those who
would have us put faith in the actual occurrence
of interruptions of that order, to produce evidence
in favour of their view, not only equal, but superior, in weight to that which leads us to adopt
ours.

This is the essential argument of Hume's
famous disquisition upon miracles; and it may
safely be declared to be irrefragable. But it must
be admitted that Hume has surrounded the kernel
of his essay with a shell of very doubtful value.

The first step in this, as in all other discussions,
is to come to a clear understanding as to the
meaning of the terms employed. Argumentation
whether miracles are possible, and, if possible,
credible, is mere beating the air until the arguers
have agreed what they mean by the word
" miracles."

Hume, with less than his usual perspicuity, but
in accordance with a common practice of believers
in the miraculous, defines a miracle as a " violation
of the laws of nature," or as " a transgression of a
law of nature by a particular volition of the Deity,
or by the interposition of some invisible agent."

There must, he says,—

"be an uniform experience against every miraculous event,
otherwise the event would not merit that appellation. And as
an uniform experience amounts to a proof, there is here a direct
and full proof, from the nature of the fact, against the existence
of any miracle ; nor can such a proof be destroyed or the miracle
rendered credible but by an opposite proof which is superior."—
(IV. p. 134.)

Every one of these dicta appears to be open to
serious objection.

The word " miracle "—*miraculum*,—in its primitive and legitimate sense, simply means something
wonderful.

Cicero applies it as readily to the fancies of
philosophers, " Portenta et miracula philosophorum
somniantium," as we do to the prodigies of priests.
And the source of the wonder which a miracle
excites is the belief, on the part of those who
witness it, that it transcends, or contradicts,
ordinary experience.

The definition of a miracle as a " violation of
the laws of nature " is, in reality, an employment
of language which, on the face of the matter,
cannot be justified. For " nature " means neither
more nor less than that which is; the sum of
phenomena presented to our experience; the
totality of events past, present, and to come.
Every event must be taken to be a part of nature
until proof to the contrary is supplied. And
such proof is, from the nature of the case, impossible.

Hume asks :—

"Why is it more than probable that all men must die : that lead cannot of itself remain suspended in the air : that fire consumes wood and is extinguished by water ; unless it be that these events are found agreeable to the laws of nature, and there is required a violation of those laws, or in other words a miracle, to prevent them ? "—(IV. p. 133.)

But the reply is obvious ; not one of these events is "more than probable " ; though the probability may reach such a very high degree that, in ordinary language, we are justified in saying that the opposite events are impossible. Calling our often verified experience a "law of nature" adds nothing to its value, nor in the slightest degree increases any probability that it will be verified again, which may arise out of the fact of its frequent verification.

If a piece of lead were to remain suspended of itself, in the air, the occurrence would be a "miracle," in the sense of a wonderful event, indeed ; but no one trained in the methods of science would imagine that any law of nature was really violated thereby. He would simply set to work to investigate the conditions under which so highly unexpected an occurrence took place ; and thereby enlarge his experience and modify his, hitherto, unduly narrow conception of the laws of nature.

The alternative definition, that a miracle is "a transgression of a law of nature by a particular volition of the Deity, or by the interposition of

some invisible agent," (IV. p. 134, *note*) is still less
defensible.   For a vast number of miracles have
professedly been worked, neither by the Deity,
nor by any invisible agent; but by Beelzebub and
his compeers, or by very visible men.

Moreover, not to repeat what has been said
respecting the absurdity of supposing that something which occurs is a transgression of laws, our
only knowledge of which is derived from the
observation of that which occurs; upon what sort
of evidence can we be justified in concluding that
a given event is the effect of a particular volition
of the Deity, or of the interposition of some
invisible (that is unperceivable) agent? It may
be so, but how is the assertion, that it is so, to be
tested? If it be said that the event exceeds the
power of natural causes, what can justify such a
saying? The day-fly has better grounds for calling a thunderstorm supernatural, than has man,
with his experience of an infinitesimal fraction of
duration, to say that the most astonishing event
that can be imagined is beyond the scope of
natural causes.

"Whatever is intelligible and can be distinctly conceived,
implies no contradiction, and can never be proved false by any
demonstration, argument, or abstract reasoning *à priori*."—(IV.
p. 44.)

So wrote Hume, with perfect justice, in his
"Sceptical Doubts." But a miracle, in the sense of
a sudden and complete change in the customary

order of nature, is intelligible, can be distinctly conceived, implies no contradiction; and therefore, according to Hume's own showing, cannot be proved false by any demonstrative argument.

Nevertheless, in diametrical contradiction to his own principles, Hume says elsewhere :—

"It is a miracle that a dead man should come to life: because that has never been observed in any age or country."— (IV. p. 134.)

That is to say, there is an uniform experience against such an event, and therefore, if it occurs, it is a violation of the laws of nature. Or, to put the argument in its naked absurdity, that which never has happened never can happen, without a violation of the laws of nature. In truth, if a dead man did come to life, the fact would be evidence, not that any law of nature had been violated, but that those laws, even when they express the results of a very long and uniform experience, are necessarily based on incomplete knowledge, and are to be held only as grounds of more or less justifiable expectation.

To sum up, the definition of a miracle as a suspension or a contravention of the order of Nature is self-contradictory, because all we know of the order of nature is derived from our observation of the course of events of which the so-called miracle is a part. On the other hand, no conceivable event, however extraordinary, is impossible; and therefore, if by the term miracles

we mean only "extremely wonderful events," there
can be no just ground for denying the possibility
of their occurrence.

But when we turn from the question of the
possibility of miracles, however they may be defined, in the abstract, to that respecting the
grounds upon which we are justified in believing
any particular miracle, Hume's arguments have a
very different value, for they resolve themselves
into a simple statement of the dictates of common
sense—which may be expressed in this canon : the
more a statement of fact conflicts with previous
experience, the more complete must be the
evidence which is to justify us in believing it. It
is upon this principle that every one carries on the
business of common life. If a man tells me he saw
a piebald horse in Piccadilly, I believe him without
hesitation. The thing itself is likely enough, and
there is no imaginable motive for his deceiving me.
But if the same person tells me he observed a zebra
there, I might hesitate a little about accepting his
testimony, unless I were well satisfied, not only
as to his previous acquaintance with zebras, but
as to his powers and opportunities of observation in the present case. If, however, my in-
formant assured me that he beheld a centaur
trotting down that famous thoroughfare, I should
emphatically decline to credit his statement ; and
this even if he were the most saintly of men and

ready to suffer martyrdom in support of his belief. In such a case, I could, of course, entertain no doubt of the good faith of the witness; it would be only his competency, which unfortunately has very little to do with good faith, or intensity of conviction, which I should presume to call in question.

Indeed, I hardly know what testimony would satisfy me of the existence of a live centaur. To put an extreme case, suppose the late Johannes Müller, of Berlin, the greatest anatomist and physiologist among my contemporaries, had barely affirmed that he had seen a live centaur, I should certainly have been staggered by the weight of an assertion coming from such an authority. But I could have got no further than a suspension of judgment. For, on the whole, it would have been more probable that even he had fallen into some error of interpretation of the facts which came under his observation, than that such an animal as a centaur really existed. And nothing short of a careful monograph, by a highly competent investigator, accompanied by figures and measurements of all the most important parts of a centaur, put forth under circumstances which could leave no doubt that falsification or misinterpretation would meet with immediate exposure, could possibly enable a man of science to feel that he acted conscientiously, in expressing his belief in the existence of a centaur on the evidence of testimony.

This hesitation about admitting the existence of such an animal as a centaur, be it observed, does not deserve reproach, as scepticism, but moderate praise, as mere scientific good faith. It need not imply, and it does not, so far as I am concerned, any *à priori* hypothesis that a centaur is an impossible animal; or, that his existence, if he did exist, would violate the laws of nature. Indubitably, the organisation of a centaur presents a variety of practical difficulties to an anatomist and physiologist; and a good many of those generalisations of our present experience, which we are pleased to call laws of nature, would be upset by the appearance of such an animal, so that we should have to frame new laws to cover our extended experience. Every wise man will admit that the possibilities of nature are infinite, and include centaurs; but he will not the less feel it his duty to hold fast, for the present, by the dictum of Lucretius, " Nam certe ex vivo Centauri non fit imago," and to cast the entire burthen of proof, that centaurs exist, on the shoulders of those who ask him to believe the statement.

Judged by the canons either of common sense, or of science, which are indeed one and the same,[1] all " miracles " are centaurs, or they would not be miracles; and men of sense and science will deal

[1] See above (p. 68) the pregnant aphorism, "philosophical decisions are nothing but the reflections of common life, methodised and corrected." [1893.]

with them on the same principles. No one who
wishes to keep well within the limits of that which
he has a right to assert will affirm that it is impossible that the sun and moon should ever have
been made to appear to stand still in the valley of
Ajalon; or that the walls of a city should have
fallen down at a trumpet blast; or that water was
turned into wine; because such events are contrary
to uniform experience and violate laws of nature.
For aught he can prove to the contrary, such events
may appear in the order of nature to-morrow.
But common sense and common honesty alike
oblige him to demand from those who would have
him believe in the actual occurrence of such events,
evidence of a cogency proportionate to their
departure from probability; evidence at least as
strong as that, which the man who says he has
seen a centaur is bound to produce, unless he is
content to be thought either more than credulous
or less than honest.

But are there any miracles on record, the
evidence for which fulfils the plain and simple
requirements alike of elementary logic and of
elementary morality?

Hume answers this question without the smallest hesitation, and with all the authority of a
historical specialist :—

"There is not to be found, in all history, any miracle attested
by a sufficient number of men, of such unquestioned goodness,
education, and learning, as to secure us against all delusion in

themselves; of such undoubted integrity, as to place them beyond all suspicion of any design to deceive others; of such credit and reputation in the eyes of mankind, as to have a great deal to lose in case of their being detected in any false-hood; and at the same time attesting facts, performed in such a public manner, and in so celebrated a part of the world, as to render the detection unavoidable: All which circumstances are requisite to give us a full assurance of the testimony of men." —(IV. p. 135.)

These are grave assertions; but they are least likely to be challenged by those who have made it their business to weigh evidence and to give their decision, under a due sense of the moral responsibility which they incur in so doing.

It is probable that few persons who proclaim their belief in miracles have considered what would be necessary to justify that belief in the case of a professed modern miracle-worker. Suppose, for example, it is affirmed that A.B. died and that C.D. brought him to life again. Let it be granted that A.B. and C.D. are persons of unimpeachable honour and veracity; that C.D. is the next heir to A.B.'s estate, and therefore had a strong motive for not bringing him to life again; and that all A.B.'s relations, respectable persons who bore him a strong affection, or had otherwise an interest in his being alive, declared that they saw him die. Furthermore, let A.B. be seen after his recovery by all his friends and neighbours, and let his and their depositions, that he is now alive, be taken down before a magistrate of known

integrity and acuteness : would all this constitute
even presumptive evidence that C.D. had worked
a miracle ? Unquestionably not. For the most
important link in the whole chain of evidence is
wanting, and that is the proof that A.B. was really
dead. The evidence of ordinary observers on such
a point as this is absolutely worthless. And, even
medical evidence, unless the physician is a person
of unusual knowledge and skill, may have little
more value. Unless careful thermometric observation proves that the temperature has sunk below
a certain point ; unless the cadaveric stiffening of
the muscles has become well established ; all the
ordinary signs of death may be fallacious, and the
intervention of C.D. may have had no more to do
with A.B.'s restoration to life than any other fortuit-
ously coincident event.

It may be said that such a coincidence would
be more wonderful than the miracle itself. Nevertheless history acquaints us with coincidences as
marvellous.

On the 19th of February, 1842, Sir Robert Sale
held Jellalabad with a small English force and,
daily expecting attack from an overwhelming
force of Afghans, had spent three months in incessantly labouring to improve the fortifications of
the town. Akbar Khan had approached within
a few miles, and an onslaught of his army was
supposed to be imminent. That morning an
earthquake—

M 2

"nearly destroyed the town, threw down the greater part of the parapets, the central gate with the adjoining bastions, and a part of the new bastion which flanked it. Three other bastions were also nearly destroyed, whilst several large breaches were made in the curtains, and the Peshawur side, eighty feet long, was quite practicable, the ditch being filled, and the descent easy. Thus, in one moment, the labours of three months were in a great measure destroyed." [1]

If Akbar Khan had happened to give orders for an assault in the early morning of the 19th of February, what good follower of the Prophet could have doubted that Allah had lent his aid? As it chanced, however, Mahometan faith in the miraculous took another turn; for the energetic defenders of the post had repaired the damage by the end of the month; and the enemy, finding no signs of the earthquake when they invested the place, ascribed the supposed immunity of Jellalabad to English witchcraft.

But the conditions of belief do not vary with time or place; and, if it is undeniable that evidence of so complete and weighty a character is needed, at the present time, for the establishment of the occurrence of such a wonder as that supposed, it has always been needful. Those who study the extant records of miracles with due attention will judge for themselves how far it has ever been supplied.

[1] Report of Captain Broadfoot, garrison engineer, quoted in Kaye's *Afghanistan*.

CHAPTER VIII

THEISM; EVOLUTION OF THEOLOGY

HUME seems to have had but two hearty dislikes : the one to the English nation, and the other to all the professors of dogmatic theology. The one aversion he vented only privately to his friends; but, if he is ever bitter in his public utterances, it is against priests[1] in general and theological enthusiasts and fanatics in particular; if he ever seems insincere, it is when he wishes to insult theologians by a parade of sarcastic respect. One need go no further than the peroration of the " Essay on Miracles " for a characteristic illustration.

[1] In a note to the Essay on Superstition and Enthusiasm, Hume is careful to define what he means by this term. "By priests I understand only the pretenders to power and dominion, and to a superior sanctity of character, distinct from virtue and good morals. These are very different from *clergymen*, who are set apart to the care of sacred matters, and the conducting our public devotions with greater decency and order. There is no rank of men more to be respected than the latter."—(III. p. 83.)

"I am the better pleased with the method of reasoning here
delivered, as I think it may serve to confound those dangerous
friends and disguised enemies to the *Christian Religion* who
have undertaken to defend it by the principles of human reason.
Our most holy religion is founded on *Faith*, not on reason, and
it is a sure method of exposing it to put it to such a trial as it is
by no means fitted to endure. . . . the Christian religion not
only was at first attended with miracles, but even at this day
cannot be believed by any reasonable person without one.
Mere reason is insufficient to convince us of its veracity: And
whoever is moved by *Faith* to assent to it, is conscious of a
continual miracle in his own person, which subverts all the
principles of his understanding, and gives him a determination
to believe what is most contrary to custom and experience."—
(IV. pp. 153, 154.)

It is obvious that, here and elsewhere, Hume,
adopting a popular confusion of ideas, uses religion
as the equivalent of dogmatic theology; and,
therefore, he says, with perfect justice, that
"religion is nothing but a species of philosophy"
(iv. p 171). Here no doubt lies the root of his
antagonism. The quarrels of theologians and
philosophers have not been about religion, but
about philosophy; and philosophers not unfrequently seem to entertain the same feeling
towards theologians that sportsmen cherish
towards poachers. "There cannot be two passions
more nearly resembling each other than hunting
and philosophy," says Hume. And philosophic
hunters are given to think, that, while they pursue
truth for its own sake, out of pure love for the
chase (perhaps mingled with a little human weak-

ness to be thought good shots), and by open and legitimate methods; their theological competitors too often care merely to supply the market of establishments; and disdain neither the aid of the snares of superstition, nor the cover of the darkness of ignorance. Unless some foundation was given for this impression by the theological writers whose works had fallen in Hume's way, it is difficult to account for the depth of feeling which so good-natured a man manifests on the subject. Thus he writes in the "Natural History of Religion," with quite unusual acerbity:

"The chief objection to it [the ancient heathen mythology] with regard to this planet is, that it is not ascertained by any just reason or authority. The ancient tradition insisted on by heathen priests and theologers is but a weak foundation : and transmitted also such a number of contradictory reports, supported all of them by equal authority, that it became absolutely impossible to fix a preference among them. A few volumes, therefore, must contain all the polemical writings of pagan priests : And their whole theology must consist more of traditional stories and superstitious practices than of philosophical argument and controversy.

"But where theism forms the fundamental principle of any popular religion, that tenet is so conformable to sound reason, that philosophy is apt to incorporate itself with such a system of theology. And if the other dogmas of that system be contained in a sacred book, such as the Alcoran, or be determined by any visible authority, like that of the Roman pontiff, speculative reasoners naturally carry on their assent, and embrace a theory, which has been instilled into them by their earliest education, and which also possesses some degree of consistence and uniformity. But as these appearances are sure,

all of them, to prove deceitful, philosophy will very soon find
herself very unequally yoked with her new associate; and
instead of regulating each principle, as they advance together,
she is at every turn perverted to serve the purposes of superstition. For besides the unavoidable incoherences, which must be
reconciled and adjusted, one may safely affirm, that all popular
theology, especially the scholastic, has a kind of appetite for
absurdity and contradiction. If that theology went not beyond
reason and common sense, her doctrines would appear too easy
and familiar.   Amazement must of necessity be raised :
Mystery affected : Darkness and obscurity sought after : And a
foundation of merit afforded to the devout votaries, who desire
an opportunity of subduing their rebellious reason by the belief
of the most unintelligible sophisms.

"Ecclesiastical history sufficiently confirms these reflections.
When a controversy is started, some people always pretend
with certainty to foretell the issue. Whichever opinion, say
they, is most contrary to plain reason is sure to prevail ; even
when the general interest of the system requires not that
decision.   Though the reproach of heresy may, for some time,
be bandied about among the disputants, it always rests at last
on the side of reason. Any one, it is pretended, that has but
learning enough of this kind to know the definition of *Arian,*
*Pelagian, Erastian, Socinian, Sabellian, Eutychian, Nestorian,
Monothelite,* &c., not to mention *Protestant,* whose fate is yet
uncertain, will be convinced of the truth of this observation.
It is thus a system becomes absurd in the end, merely from its
being reasonable and philosophical in the beginning.

"To oppose the torrent of scholastic religion by such feeble
maxims as these, that *it is impossible for the same thing to be
and not to be,* that *the whole is greater than a part,* that *two and
three make five,* is pretending to stop the ocean with a bulrush.
Will you set up profane reason against sacred mystery ? No
punishment is great enough for your impiety. And the same
fires which were kindled for heretics will serve also for the
destruction of philosophers."—(IV. pp. 481—3.)

Holding these opinions respecting the recognised

systems of theology and their professors, Hume, nevertheless, seems to have had a theology of his own; that is to say, he seems to have thought (though, as will appear, it is needful for an expositor of his opinions to speak very guardedly on this point) that the problem of theism is suscept- ible of scientific treatment, with something more than a negative result. His opinions are to be gathered from the eleventh section of the "Inquiry" (1748); from the "Dialogues concerning Natural Religion," which were written at least as early as 1751, though not published till after his death; and from the "Natural History of Religion," pub- lished in 1757.

In the first two pieces, the reader is left to judge for himself which interlocutor in the dialogue represents the thoughts of the author; but for the views put forward in the last, Hume accepts the responsibility. Unfortunately, this essay deals almost wholly with the historical development of theological ideas; and, on the question of the philosophical foundation of theology, does little more than express the writer's contentment with the argument from design.

"The whole frame of nature bespeaks an Intelligent Author; and no rational inquirer can, after serious reflection, suspend his belief a moment with regard to the primary prin- ciples of genuine Theism and Religion.—(IV. p. 435.)

"Were men led into the apprehension of invisible, intel- ligent power, by a contemplation of the works of nature, they could never possibly entertain any conception but of one

single being, who bestowed existence and order on this vast
machine, and adjusted all its parts according to one regular
plan or connected system.   For though, to persons of a
certain turn of mind, it may not appear altogether absurd,
that several independent beings, endowed with superior
wisdom, might conspire in the contrivance and execution of
one regular plan, yet is this a merely arbitrary supposition,
which, even if allowed possible, must be confessed neither to
be supported by probability nor necessity.   All things in the
universe are evidently of a piece.   Everything is adjusted to
everything.   One design prevails throughout the whole.   And
this uniformity leads the mind to acknowledge one author ;
because the conception of different authors, without any distinction of attributes or operations, serves only to give perplexity to the imagination, without bestowing any satisfaction
on the understanding."—(IV. p. 442.)

Thus Hume appears to have sincerely accepted
the two fundamental conclusions of the argument
from design; firstly, that a Deity exists; and,
secondly, that He possesses attributes more or less
allied to those of human intelligence.   But, at this
embryonic stage of theology, Hume's progress is
arrested; and, after a survey of the development
of dogma, his " general corollary " is that—

" The whole is a riddle, an enigma, an inexplicable mystery.
Doubt, uncertainty, suspense of judgment, appear the only
result of our most accurate scrutiny concerning this subject.
But such is the frailty of human reason and such the irre-
sistible contagion of opinion, that even this deliberate doubt
could scarcely be upheld ; did we not enlarge our view, and
opposing one species of superstition to another, set them a
quarrelling ; while we ourselves, during their fury and con-
tention, happily make our escape into the calm, though obscure,
regions of philosophy."—(IV. p. 513.)

Thus it may be fairly presumed that Hume expresses his own sentiments in the words of the
speech with which Philo concludes the "Dialogues."

"If the whole of natural theology, as some people seem to
maintain, resolves itself into one simple, though somewhat
ambiguous, at least undefined proposition, *That the cause or
causes of order in the universe probably bear some remote analogy
to human intelligence:* If this proposition be not capable of
extension, variation, or more particular explication : If it
affords no inference that affects human life or can be the
source of any action or forbearance : And if the analogy,
imperfect as it is, can be carried no further than to the human
intelligence, and cannot be transferred, with any appearance
of probability, to the other qualities of the mind ; if this really
be the case, what can the most inquisitive, contemplative,
and religious man do more than give a plain, philosophical
assent to the proposition, as often as it occurs, and believe
that the arguments on which it is established exceed the
objections which lie against it ? Some astonishment indeed
will naturally arise from the greatness of the object ; some
melancholy from its obscurity ; some contempt of human
reason, that it can give no solution more satisfactory with
regard to so extraordinary and magnificent a question.   But
believe me, Cleanthes, the most natural sentiment which a
well-disposed mind will feel on this occasion, is a longing
desire and expectation that Heaven would be pleased to dis-
sipate, at least alleviate, this profound ignorance, by affording
some more particular revelation to mankind, and making
discoveries of the nature, attributes, and operations of the
Divine object of our faith."[1]—(II. pp. 547—8.)

---

[1] It is needless to quote the rest of the passage, though I cannot
refrain from observing that the recommendation which it contains
that a "man of letters" should become a philosophical sceptic as
"the first and most essential step towards being a sound believing
Christian," though adopted and largely acted upon by many a
champion of orthodoxy in these days, is questionable in taste, if it

Such being the sum total of Hume's conclusions it cannot be said that his theological burden is a heavy one. But, if we turn from the "Natural History of Religion," to the "Treatise," the "Inquiry," and the "Dialogues," the story of what happened to the ass laden with salt, who took to the water, irresistibly suggests itself. Hume's theism, such as it is, dissolves away in the dialectic river, until nothing is left but the verbal sack in which it was contained.

Of the two theistic propositions to which Hume is committed, the first is the affirmation of the existence of a God, supported by the argument from the nature of causation. In the "Dialogues," Philo, while pushing scepticism to its utmost limit, is nevertheless made to say that—

". . . . where reasonable men treat these subjects, the question can never be concerning the *Being*, but only the *Nature* of the Deity. The former truth, as you will observe, is unquestionable and self-evident. Nothing exists without a cause, and the original cause of this universe (whatever it be) we call God, and piously ascribe to him every species of perfection."—(II. p. 439.)

The expositor of Hume, who wishes to do his work thoroughly, as far as it goes, cannot but fall

be meant as a jest, and more than questionable in morality, if it is to be taken in earnest. To pretend that you believe any doctrine for no better reason than that you doubt everything else, would be dishonest, if it were not preposterous.

into perplexity [1] when he contrasts this language
with that of the sections of the third part of the
"Treatise," entitled, *Why a Cause is Always Necessary* and *Of the Idea of Necessary Connexion*.

It is there shown at large that, "every demonstration which has been produced for the necessity of a
cause is fallacious and sophistical" (I. p. 111) ; it
is affirmed, that "there is no absolute nor metaphysical necessity that every beginning of existence
should be attended with such an object" [as a
cause] (I. p. 227) ; and it is roundly asserted, that

[1] A perplexity which is increased rather than diminished by
some passages in a letter to Gilbert Elliot of Minto (March 10,
1751). Hume says, "You would perceive by the sample I
have given you that I make Cleanthes the hero of the dialogue ;
whatever you can think of, to strengthen that side of the argument, will be most acceptable to me. Any propensity you
imagine I have to the other side crept in upon me against my
will ; and 'tis not long ago that I burned an old manuscript
book, wrote before I was twenty, which contained, page after
page, the gradual progress of my thoughts on this head. It
began with an anxious scent after arguments to confirm the
common opinion ; doubts stole in, dissipated, returned ; were
again dissipated, returned again ; and it was a perpetual struggle
of a restless imagination against inclination—perhaps against
reason. . . . I could wish Cleanthes' argument could be so
analysed as to be rendered quite formal and regular. The propensity of the mind towards it—unless that propensity were as
strong and universal as that to believe in our senses and experience—will still, I am afraid, be esteemed a suspicious foundation. 'Tis here I wish for your assistance. We must endeavour
to prove that this propensity is somewhat different from our
inclination to find our own figures in the clouds, our faces in the
moon, our passions and sentiments even in inanimate matter.
Such an inclination may and ought to be controlled, and can never
be a legitimate ground of assent." (Burton, *Life*, I. pp. 331—
3.) The picture of Hume here drawn unconsciously by his own
hand, is unlike enough to the popular conception of him as a
careless sceptic, loving doubt for doubt's sake.

it is " easy for us to conceive any object to be non-
existent this moment and existent the next, without conjoining to it the distinct idea of a cause or
productive principle " (I. p. 111). So far from the
axiom, that whatever begins to exist must have a
cause of existence, being "self-evident," as Philo
calls it, Hume spends the greatest care in showing
that it is nothing but the product of custom, or
experience.

And the doubt thus forced upon one, whether
Philo ought to be taken as Hume's mouthpiece
even so ' far, is increased when we reflect that we
are dealing with an acute reasoner; and that
there is no difficulty in drawing the deduction
from Hume's own definition of a cause, that the
very phrase, a " first cause," involves a contradiction in terms. He lays down that,—

" 'Tis an established axiom both in natural and moral philosophy, that an object, which exists for any time in its full
perfection without producing another, is not its sole cause ; but
is assisted by some other principle which pushes it from its state
of inactivity, and makes it exert that energy, of which it was
secretly possessed."—(I. p. 106.)

Now the " first cause " is assumed to have existed from all eternity, up to the moment at which
the universe came into existence.   Hence it cannot
be the sole cause of the universe; in fact, it was
no cause at all until it was " assisted by some
other principle "; consequently the so-called
" first cause," so far as it produces the universe,

is in reality an effect of that other principle. Moreover, though, in the person of Philo, Hume assumes the axiom "that whatever begins to exist must have a cause," which he denies in the " Treatise," he must have seen, for a child may see, that the assumption is of no real service.

Suppose Y to be the imagined first cause and Z to be its effect. Let the letters of the alphabet, a, b, c, d, e, f, g, in their order, represent successive moments of time, and let g represent the particular moment at which the effect Z makes its appearance. It follows that the cause Y could not have existed "in its full perfection" during the time $a—e$, for if it had, then the effect Z would have come into existence during that time, which, by the hypothesis, it did not do. The cause Y, therefore, must have come into existence at f, and if " everything that comes into exitsence has a cause," Y must have had a cause X operating at e, X a cause W operating at d; and so on, ad infinitum.[1]

If the only demonstrative argument for the existence of a Deity, which Hume advances, thus literally, "goes to water" in the solvent of his philosophy, the reasoning from the evidence of design does not fare much better. If Hume really

[1] Kant employs substantially the same argument :—" Würde das höchste Wesen in dieser Kette der Bedingungen stehen, so würde es selbst ein Glied der Reihe derselben sein, und eben so wie die niederen Glieder, denen es vorgesetzt ist, noch fernere Untersuchungen wegen seines noch höheren Grundes erfahren." —*Kritik.* Ed. Hartenstein, p. 422

knew of any valid reply to Philo's arguments in
the following passages of the "Dialogues," he has
dealt unfairly by the reader in concealing it :—

"But because I know you are not much swayed by names
and authorities, I shall endeavour to show you, a little more
distinctly, the inconveniences of that Anthropomorphism, which
you have embraced ; and shall prove that there is no ground
to suppose a plan of the world to be formed in the Divine
mind, consisting of distinct ideas, differently arranged, in the
same manner as an architect forms in his head the plan of a
house which he intends to execute.

"It is not easy, I own, to see what is gained by this supposition, whether we judge the matter by *Reason* or by *Experience.* We are still obliged to mount higher in order to find
the cause of this cause, which you had assigned as satisfactory
and conclusive.

"If *Reason* (I mean abstract reason, derived from inquiries à
priori) be not alike mute with regard to all questions concerning cause and effect, this sentence at least it will venture to
pronounce : That a mental world, or universe of ideas, requires
a cause as much as does a material world or universe of
objects ; and, if similar in its arrangement, must require a
similar cause. For what is there in this subject, which should
occasion a different conclusion or inference ? In an abstract
view they are entirely alike ; and no difficulty attends the one
supposition, which is not common to both of them.

"Again, when we will needs force *Experience* to pronounce
some sentence, even on those subjects which lie beyond her
sphere, neither can she perceive any material difference in this
particular, between these two kinds of worlds ; but finds them
to be governed by similar principles, and to depend upon an
equal variety of causes in their operations. We have specimens
in miniature of both of them. Our own mind resembles the
one ; a vegetable or animal body the other. Let experience,
therefore, judge from these samples. Nothing seems more
delicate, with regard to its causes, than thought : and as these
causes never operate in two persons after the same manner, so

we never find two persons who think exactly alike. Nor indeed does the same person think exactly alike at any two different periods of time. A difference of age, of the disposition of his body, of weather, of food, of company, of books, of passions; any of these particulars, or others more minute, are sufficient to alter the curious machinery of thought, and communicate to it very different movements and operations. As far as we can judge, vegetables and animal bodies are not more delicate in their motions, nor depend upon a greater variety or more curious adjustment of springs and principles.

"How, therefore, shall we satisfy ourselves concerning the cause of that Being whom you suppose the Author of Nature, or, according to your system of anthropomoirphsm, the ideal world in which you trace the material? Have we not the same reason to trace the ideal world into another ideal world, or new intelligent principle? But if we stop and go no farther; why go so far? Why not stop at the material world? How can we satisfy ourselves without going on in infinitum? And after all, what satisfaction is there in that infinite progression? Let us remember the story of the Indian philosopher and his elephant. It was never more applicable than to the present subject. If the material world rests upon a similar ideal world, this ideal world must rest upon some other; and so on without end. It were better, therefore, never to look beyond the present material world. By supposing it to contain the principle of its order within itself, we really assert it to be God; and the sooner we arrive at that Divine Being, so much the better. When you go one step beyond the mundane system you only excite an inquisitive humour, which it is impossible ever to satisfy.

"To say, that the different ideas which compose the reason of the Supreme Being, fall into order of themselves and by their own natures, is really to talk without any precise meaning. If it has a meaning, I would fain know why it is not as good sense to say, that the parts of the material world fall into order of themselves, and by their own nature. Can the one opinion be intelligible while the other is not so?"
—(II. pp. 461—4.)

Cleanthes, in replying to Philo's discourse, says that it is very easy to answer his arguments; but, as not unfrequently happens with controversialists, he mistakes a reply for an answer, when he declares that—

"The order and arrangement of nature, the curious adjustment of final causes, the plain use and intention of every part and organ; all these bespeak in the clearest language one intelligent cause or author. The heavens and the earth join in the same testimony. The whole chorus of nature raises one hymn to the praises of its Creator."—(II. p. 465.)

Though the rhetoric of Cleanthes may be admired, its irrelevancy to the point at issue must be admitted. Wandering still further into the region of declamation, he works himself into a passion :

"You alone, or almost alone, disturb this general harmony. You start abstruse doubts, cavils, and objections : You ask me what is the cause of this cause? I know not: I care not: that concerns not me. I have found a Deity; and here I stop my inquiry. Let those go further who are wiser or more enterprising."—(II. p. 466.)

In other words, O Cleanthes, reasoning having taken you as far as you want to go, you decline to advance any further; even though you fully admit that the very same reasoning forbids you to stop where you are pleased to cry halt! But this is simply forcing your reason to abdicate in favour of your caprice. It is impossible to imagine that Hume, of all men in the world, could have rested satisfied with such an act of high-treason against the sovereignty of philosophy. We may rather conclude that the last word of the discussion, which he gives to Philo, is also his own.

" If I am still to remain in utter ignorance of causes, and can absolutely give an explication of nothing, I shall never esteem it any advantage to shove off for a moment a difficulty, which, you acknowledge, must immediately, in its full force, recur upon me. Naturalists [1] indeed very justly explain particular effects by more general causes, though these general causes should remain in the end totally inexplicable ; but they never surely thought it satisfactory to explain a particular effect by a particular cause, which was no more to be accounted for than the effect itself. An ideal system, arranged of itself, without a precedent design, is not a whit more explicable than a material one, which attains its order in a like manner ; nor is there any more difficulty in the latter supposition than in the former."—(II. p. 466.)

It is obvious that, if Hume had been pushed, he must have admitted that his opinion concerning the existence of a God, and of a certain remote resemblance of his intellectual nature to that of man, was an hypothesis which might possess more or less probability, but, on his own principles, was incapable of any approach to demonstration. And to all attempts to make any practical use of his theism; or to prove the existence of the attributes of infinite wisdom, benevolence, justice, and the like, which are usually ascribed to the

---

[1] *I.e.* Natural philosophers.

Deity, by reason, he opposes a searching critical negation.[1]

The object of the speech of the imaginary Epicurean in the eleventh section of the "Inquiry," entitled "Of a Particular Providence and of a Future State," is to invert the argument of Bishop Butler's "Analogy."

That famous defence of theology against the *a priori* scepticism of Freethinkers of the eighteenth century, who based their arguments on the inconsistency of the revealed scheme of salvation with the attributes of the Deity, consists, essentially, in conclusively proving that, from a moral point of view, Nature is at least as reprehensible as orthodoxy. If you tell me, says Butler, in effect, that any part of revealed religion must be false because it is inconsistent with the divine attributes of justice and mercy; I beg leave to point out to you, that there are undeniable natural facts which are fully open to the same objection. Since you admit that nature is the work of God, you are forced to allow that such facts are consistent with his attributes. Therefore, you must also admit, that the parallel facts in the scheme of orthodoxy are also consistent with them, and all your arguments to the contrary fall to the ground. Q.E.D. In fact, the

---

[1] Hume's letter to Mure of Caldwell, containing a criticism of Leechman's sermon (Burton, I. p. 163), bears strongly on this point.

solid sense of Butler left the Deism of the Freethinkers not a leg to stand upon. Perhaps, however, he did not remember the wise saying that " A man seemeth right in his own cause, but another cometh after and judgeth him." Hume's Epicurean philosopher adopts the main arguments of the " Analogy," but unfortunately drives them home to a conclusion of which the good Bishop would hardly have approved.

"I deny a Providence, you say, and supreme governor of the world, who guides the course of events, and punishes the vicious with infamy and disappointment, and rewards the virtuous with honour and success in all their undertakings. But surely I deny not the course itself of events which lies open to every one's inquiry and examination. I acknowledge that, in the present order of things, virtue is attended with more peace of mind than vice, and meets with a more favour able reception from the world. I am sensible that, according to the past experience of mankind, friendship is the chief joy of human life, and moderation the only source of tranquillity and happiness. I never balance between the virtuous and the vicious course of life; but am sensible that, to a well-disposed mind, every advantage is on the side of the former. And what can you say more, allowing all your suppositions and reasonings ? You tell me, indeed, that this disposition of things proceeds from intelligence and design. But, whatever it proceeds from, the disposition itself, on which depends our happiness and misery, and consequently our conduct and deportment in life, is still the same. It is still open for me, as well as you, to regulate my behaviour by my experience of past events. And if you affirm that, while a divine providence is allowed, and a supreme distributive justice in the universe, I ought to expect some more particular reward of the good, and pun-ishment of the bad, beyond the ordinary course of events, I here find the same fallacy which I have before endeavoured to detect. You persist in imagining, that if we grant that divine existence for which you so earnestly contend, you may safely infer consequences from it, and add something to the experienced order of nature by arguing from the attributes which you ascribe to your gods. You seem not to remember that all your reasonings on this subject can only be drawn from effects to causes ; and that every argument, deduced from causes to effects, must of necessity be a gross sophism, since it is impossible for you to know anything of the cause, but what you have antecedently not inferred, but discovered to the full, in the effect.

"But what must a philosopher think of those vain reasoners who, instead of regarding the present scene of things as the sole object of their contemplation, so far reverse the whole course of nature, as to render this life merely a passage to something further; a porch, which leads to a greater and vastly different building; a prologue which serves only to introduce the piece, and give it more grace and propriety? Whence, do you think, can such philosophers derive their idea of the gods? From their own conceit and imagination surely. For if they derive it from the present phenomena, it would never point to anything further, but must be exactly adjusted to them. That the divinity may *possibly* be endowed with attributes which we have never seen exerted ; may be governed by principles of action which we cannot discover to be satisfied'; all this will freely be allowed. But still this is mere *possibility* and hypothesis. We never can have reason to *infer* any attributes or any principles of action in him, but so far as we know them to have been exerted and satisfied.

"*Are there any marks of a distributive justice in the world?* If you answer in the affirmative, I conclude that since justice here exerts itself, it is satisfied. If you reply in the negative, I conclude that you have then no reason to ascribe justice, in our sense of it, to the gods. If you hold a medium between affirmation and negation, by saying that the justice of the gods at present exerts itself in part, but not in its full extent, I answer that you have no reason to give it any particular extent, but only so far as you see it, at *present*, exert itself."
(IV pp. 164—6.)

Thus, the Freethinkers said, the attributes of the Deity being what they are, the scheme of orthodoxy is inconsistent with them; whereupon Butler gave the crushing reply: Agreeing with you as to the attributes of the Deity, nature, by its existence, proves that the things to which you object are quite consistent with them. To whom enters Hume's Epicurean with the remark : Then, as nature is our only measure of the attributes of the Deity in their practical manifestation, what warranty is there for supposing that such measure is anywhere transcended ? That the "other side" of nature, if there be one, is governed on different principles from this side ?

Truly on this topic silence is golden; while speech reaches not even the dignity of sounding brass or tinkling cymbal, and is but the weary clatter of an endless logomachy. One can but suspect that Hume also had reached this conviction; and that his shadowy and inconsistent theism was the expression of his desire to rest in a state of mind, which distinctly excluded negation, while it included as little as possible of affirmation, respecting a problem which he felt to be hopelessly insoluble.

But, whatever might be the views of the philosopher as to the arguments for theism, the historian could have no doubt respecting its many-shaped existence, and the great part which it has played in the world. Here, then, was a

body of natural facts to be investigated scientifically, and the result of Hume's inquiries is
embodied in the remarkable essay on the
"Natural History of Religion." Hume anticipated the results of modern investigation in
declaring fetishism and polytheism to be the
form in which savage and ignorant men naturally
clothe their ideas of the unknown influences
which govern their destiny; and they are polytheists rather than monotheists because,—

". . . The first ideas of religion arose, not from a contemplation of the works of nature, but from a concern with regard
to the events of life, and from the incessant hopes and fears
which actuate the human mind. . . . in order to carry men's
attention beyond the present course of things, or lead them
into any inference concerning invisible intelligent power, they
must be actuated by some passion which prompts their thought
and reflection, some motive which urges their first enquiry.
But what passion shall we have recourse to, for explaining an
effect of such mighty consequence ? Not speculative curiosity
merely, or the pure love of truth. That motive is too refined
for such gross apprehensions, and would lead men into enquiries
concerning the frame of nature, a subject too large and compre-
hensive for their narrow capacities. No passions, therefore, can
be supposed to work on such barbarians, but the ordinary affections of human life ; the anxious concern for happiness, the
dread of future misery, the terror of death, the thirst of re-
venge, the appetite for food and other necessaries. Agitated by
hopes and fears of this nature, especially the latter, men scru-
tinize, with a trembling curiosity, the course of future causes,
and examine the various and contrary events of human life.
And in this disordered scene, with eyes still more disordered
and astonished, they see the first obscure traces of divinity."—
(IV. pp. 443—4.)

The shape assumed by these first traces of divinity is that of the shadows of men's own minds, projected out of themselves by their imaginations :—

"There is an universal tendency among mankind to conceive all beings like themselves, and to transfer to every object those qualities with which they are familiarly acquainted, and of which they are intimately conscious. . . The *unknown causes* which continually employ their thought, appearing always in the same aspect, are all apprehended to be of the same kind or species. Nor is it long before we ascribe to them thought, and reason, and passion, and sometimes even the limbs and figures of men in order to bring them nearer to a resemblance with ourselves."—(IV. pp. 446—7.)

Hume asks whether polytheism really deserves the name of theism.

"Our ancestors in Europe, before the revival of letters, believed as we do at present, that there was one supreme God, the author of nature, whose power, though in itself uncontrollable, was yet often exerted by the interposition of his angels and subordinate ministers, who executed his sacred purposes. But they also believed, that all nature was full of other invisible powers: fairies, goblins, elves, sprights ; beings stronger and mightier than men, but much inferior to the celestial natures who surround the throne of God. Now, suppose that any one, in these ages, had denied the existence of God and of his angels, would not his impiety justly have deserved the appellation of atheism, even though he had still allowed, by some odd capricious reasoning, that the popular stories of elves and fairies were just and well grounded? The difference, on the one hand, between such a person and a genuine theist, is infinitely greater than that, on the other, between him and one that absolutely excludes all invisible intelligent power. And it is a fallacy, merely from the casual resemblance of names, without any

conformity of meaning, to rank such opposite opinions under
the same denomination.

"To any one who considers justly of the matter, it will
appear that the gods of the polytheists are no better than the
elves and fairies of our ancestors, and merit as little as any pious
worship and veneration. These pretended religionists are really
a kind of superstitious atheists, and acknowledge no being that
corresponds to our idea of a Deity. No first principle of mind
or thought; no supreme government and administration; no
divine contrivance or intention in the fabric of the world."—
(IV. pp. 450—51.)

The doctrine that you may call an atheist
anybody whose ideas about the Deity do not
correspond with your own, is so largely acted
upon by persons who are certainly not of Hume's
way of thinking and, probably, so far from having
read him, would shudder to open any book
bearing his name, except the "History of England,"
that it is surprising to trace the theory of their
practice to such a source.

But on thinking the matter over, this theory
seems so consonant with reason, that one feels
ashamed of having suspected many excellent
persons of being moved by mere malice and
viciousness of temper to call other folks atheists,
when, after all, they have been obeying a purely
intellectual sense of fitness. As Hume says, truly
enough, it is a mere fallacy, because two people
use the same names for things, the ideas of which
are mutually exclusive, to rank such opposite
opinions under the same denomination. If the

Jew says, that the Deity is absolute unity, and that it is sheer blasphemy to say that He ever became incarnate in the person of a man; and, if the Trinitarian says, that the Deity is numerically three as well as numerically one, and that it is sheer blasphemy to say that He did not so become incarnate, it is obvious enough that each must be logically held to deny the existence of the other's Deity.   Therefore; that each has a scientific right to call the other an atheist; and that, if he refrains, it is only on the ground of decency and good manners, which should restrain an honourable man from employing even scientifically justifiable language, if custom has given it an abusive connotation.   While one must agree with Hume, then, it is, nevertheless, to be wished that he had not set the bad example of calling polytheists "superstitious atheists."   It probably did not occur to him that, by a parity of reasoning, the Unitarians might justify the application of the same language to the Ultramontanes, and *vice versâ.*   But, to return from a digression which may not be wholly unprofitable, Hume proceeds to show in what manner polytheism incorporated physical and moral allegories, and naturally accepted hero-worship; and he sums up his views of the first stages of the evolution of theology as follows:—

"These then are the general principles of polytheism, founded in human nature, and little or nothing dependent on caprice or

accident. As the *causes* which bestow happiness or misery, are
in general very little known and very uncertain, our anxious
concern endeavours to attain a determinate idea of them ; and
finds no better expedient than to represent them as intelligent,
voluntary agents, like ourselves, only somewhat superior in
power and wisdom. The limited influence of these agents, and
their proximity to human weakness, introduce the various
distribution and division of their authority, and thereby give
rise to allegory. The same principles naturally deify mortals,
superior in power, courage, or understanding, and produce hero-
worship ; together with fabulous history and mythological
tradition, in all its wild and unaccountable forms. And as an
invisible spiritual intelligence is an object too refined for vulgar
apprehension, men naturally affix it to some sensible representation ; such as either the more conspicuous parts of nature, or
the statues, images, and pictures, which a more refined age
forms of its divinities."—(IV. p. 461.)

How did the further stage of theology, monotheism, arise out of polytheism ? Hume replies,
certainly not by reasonings from first causes or
any sort of fine-drawn logic :—

"Even at this day, and in Europe, ask any of the vulgar why
he believes in an Omnipotent Creator of the world, he will
never mention the beauty of final causes, of which he is wholly
ignorant : He will not hold out his hand and bid you contem-
plate the suppleness and variety of joints in his fingers, their
bending all one way, the counterpoise which they receive from
the thumb, the softness and fleshy parts of the inside of the
hand, with all the other circumstances which render that
member fit for the use to which it was destined. To these he has
been long accustomed ; and he beholds them with listlessness and
unconcern. He will tell you of the sudden and unexpected death
of such-a-one ; the fall and bruise of such another ; the excessive
drought of this season ; the cold and rains of another. These he
ascribes to the immediate operation of Providence : And such

events as, with good reasoners, are the chief difficulties in admitting a Supreme Intelligence, are with him the sole arguments for
it. . . .

"We may conclude therefore, upon the whole, that since the
vulgar, in nations which have embraced the doctrine of theism,
still build it upon irrational and superstitious grounds, they are
never led into that opinion by any process of argument, but by
a certain train of thinking, more suitable to their genius and
capacity.

"It may readily happen, in an idolatrous nation, that though
men admit the existence of several limited deities, yet there is
some one God, whom, in a particular manner, they make the
object of their worship and adoration.   They may either suppose, that, in the distribution of power and territory among the
Gods, their nation was subjected to the jurisdiction of that
particular deity ; or, reducing heavenly objects to the model of
things below, they may represent one god as the prince or
supreme magistrate of the rest, who, though of the same nature,
rules them with an authority like that which an earthly sovereign exerts over his subjects and vassals.   Whether this god,
therefore, be considered as their peculiar patron, or as the
general sovereign of heaven, his votaries will endeavour, by
every art, to insinuate themselves into his favour ; and supposing him to be pleased, like themselves, with praise and flattery,
there is no eulogy or exaggeration which will be spared in their
addresses to him.   In proportion as men's fears or distresses
become more urgent, they still invent new strains of adulation ;
and even he who outdoes his predecessor in swelling the titles
of his divinity, is sure to be outdone by his successor in newer
and more pompous epithets of praise.   Thus they proceed, till
at last they arrive at infinity itself, beyond which there is no
further progress ; And it is well if, in striving to get further,
and to represent a magnificent simplicity, they run not into
inexplicable mystery, and destroy the intelligent nature of their
deity, on which alone any rational worship or adoration can be
founded.   While they confine themselves to the notion of a
perfect being, the Creator of the world, they coincide, by chance,
with the principles of reason and true philosophy ; though they

are guided to that notion, not by reason, of which they are in a
great measure incapable, but by the adulation and fears of the
most vulgar superstition."—(IV. pp. 463-6.)

"Nay, if we should suppose, what never happens, that a
popular religion were found, in which it was expressly declared,
that nothing but morality could gain the divine favour; if
an order of priests were instituted to inculcate this opinion,
in daily sermons, and with all the arts of persuasion; yet so
inveterate are the people's prejudices, that, for want of some
other superstition they would make the very attendance on
these sermons the essentials of religion, rather than place
them in virtue and good morals. The sublime prologue of
Zaleucus' laws inspired not the Locrians, so far as we can
learn, with any sounder notions of the measures of acceptance
with the deity, than were familiar to the other Greeks."—
(IV. p. 505.)

It has been remarked that Hume's writings are
singularly devoid of local colour; of allusions to
the scenes with which he was familiar, and to the
people from whom he sprang. Yet, surely, the
Lowlands of Scotland were more in his thoughts
than the Zephyrean promontory, and the hard
visage of John Knox peered from behind the
mask of Zaleucus, when this passage left his pen.
Nay, might not an acute German critic discern
therein a reminiscence of that eminently Scottish
institution, a "Holy Fair"? where, as Hume's
young contemporary sings :—

> "* * * opens out his cauld harangues
>   On practice and on morals ;
> An' aff the godly pour in thrangs
>   To gie the jars and barrels
>   A lift that day.

" What signifies his barren shine
    Of moral powers and reason ?
His English style and gesture fine
    Are a' clean out of season.
Like Socrates or Antonine,
    Or some auld pagan heathen,
The moral man he does define,
    But ne'er a word o' faith in
        That's right that day." [1]

---

[1] Burns published the *Holy Fair* only ten years after Hume's death.

CHAPTER IX

DESCARTES taught that an absolute difference of kind separates matter, as that which possesses extension, from spirit, as that which thinks. They not only have no character in common, but it is inconceivable that they should have any. On the assumption, that the attributes of the two were wholly different, it appeared to be a necessary consequence that the hypothetical causes of these attributes—their respective substances—must be totally different. Notably, in the matter of divisibility, since that which has no extension cannot be divisible, it seemed that the *chose pensante*, the soul, must be an indivisible entity.

Later philosophers, accepting this notion of the soul, were naturally much perplexed to understand how, if matter and spirit had nothing in common, they could act and react on one another. All the changes of matter being modes of motion, the difficulty of understanding how a moving extended material body was to affect a thinking thing which had no dimension, was as great as that involved in solving the problem of how to hit a nominative case with a stick. Hence, the successors of Descartes either found themselves obliged, with the Occasionalists, to call in the aid of the Deity, who was supposed to be a sort of go-between betwixt matter and spirit; or they had recourse, with Leibnitz, to the doctrine of pre-established harmony, which denied any influence of the body on the soul, or *vice versâ*, and compared matter and spirit to two clocks so accurately regulated to keep time with one another, that the one struck whenever the other pointed to the hour; or, with Berkeley, they abolished the "substance" of matter altogether, as a superfluity, though they failed to see that the same arguments equally justified the abolition of soul as another superfluity, and the reduction of the universe to a series of events or phenomena; or, finally, with Spinoza, to whom Berkeley makes a perilously close approach, they asserted the existence of only one substance, with two chief attributes, the one, thought, and the other, extension.

There remained only one possible position, which, had it been taken up earlier, might have saved an immensity of trouble; and that was to affirm that we do not, and cannot, know anything about the

"substance" either of the thinking thing, or of
the extended thing. And Hume's sound common
sense led him to defend the thesis which Locke
had already foreshadowed, with respect to the
question of the substance of the soul. Hume
enunciates two opinions. The first is that the
question itself is unintelligible, and therefore
cannot receive any answer; the second is that
the popular doctrine respecting the immateriality,
simplicity, and indivisibility of a thinking substance is a " true atheism, and will serve to justify
all those sentiments for which Spinoza is so
universally infamous."

In support of the first opinion, Hume points out
that it is impossible to attach any definite meaning to the word " substance" when employed for
the hypothetical substratum of soul and matter.
For if we define substance as that which may
exist by itself, the definition does not distinguish
the soul from perceptions. It is perfectly easy to
conceive that states of consciousness are self-sub-
sistent. And, if the substance of the soul is
defined as that in which perceptions inhere, what
is meant by the inherence ? Is such inherence
conceivable ? If conceivable, what evidence is
there of it ? And what is the use of a substratum
to things which, for anything we know to the
contrary, are capable of existing by themselves ?

Moreover, it may be added, supposing the soul
has a substance, how do we know that it is differ-

ent from the substance, which, on like grounds, must be supposed to underlie the qualities of matter?

Again, if it be said that our personal identity requires the assumption of a substance which remains the same while the accidents of perception shift and change, the question arises what is meant by personal identity?

"For my part," says Hume, "when I enter most intimately into what I call *myself*, I always stumble on some particular perception or other, of heat or cold, light or shade, love or hatred, pain or pleasure. I never can catch *myself* at any time without a perception, and never can observe anything but the perception. When my perceptions are removed for any time, as by sound sleep, so long am I insensible of *myself*, and may be truly said not to exist. And were all my perceptions removed by death, and I could neither think, nor feel, nor see, nor love, nor hate, after the dissolution of my body, I should be entirely annihilated, nor do I conceive what is further requisite to make me a perfect nonentity. If any one, upon serious and unprejudiced reflection, thinks he has a different notion of *himself*, I must confess I can reason no longer with him. All I can allow him is, that he may be in the right as well as I, and that we are essentially different in this particular. He may perhaps perceive something simple and continued which he calls *himself*, though I am certain there is no such principle in me.

"But setting aside some metaphysicians of this kind, I may venture to affirm of the rest of mankind, that they are nothing but a bundle or collection of different perceptions, which succeed one another with an inconceivable rapidity, and are in a perpetual flux and movement. . . . The mind is a kind of theatre, where several perceptions successively make their appearance, pass, repass, glide away, and mingle in an infinite variety of postures and situations. There is properly no

o 2

simplicity in it at one time, nor *identity* in different, whatever
natural propension we may have to imagine that simplicity
and identity. The comparison of the theatre must not mislead
us. They are the successive perceptions only that constitute
the mind ; nor have we the most distant notion of the place
where these scenes are represented, or of the materials of which
it is composed.

"What then gives so great a propension to ascribe an
identity to these successive perceptions, and to suppose ourselves possessed of an invariable and uninterrupted existence
through the whole course of our lives ? In order to answer
this question, we must distinguish between personal identity
as it regards our thought and imagination, and as it regards
our passions, or the concern we take in ourselves. The first
is our present subject ; and to explain it perfectly we must
take the matter pretty deep, and account for that identity
which we attribute to plants and animals ; there being a great
analogy betwixt it and the identity of a self or person."—(I.
pp. 321, 322.)

Perfect identity is exhibited by an object
which remains unchanged throughout a certain
time; perfect diversity is seen in two or more
objects which are separated by intervals of space
and periods of time. But, in both these cases,
there is no sharp line of demarcation between
identity and diversity, and it is impossible to say
when an object ceases to be one and becomes
two.

When a sea-anemone multiplies, by division,
there is a time during which it is said to be one
animal partially divided; but after a while, it
becomes two animals adherent together, and the
limit between these conditions is purely arbitrary.

So in mineralogy, a crystal of a definite chemical composition may have its substance replaced, particle by particle, by another chemical compound. When does it lose its primitive identity and become a new thing?

Again, a plant or an animal, in the course of its existence, from the condition of an egg or seed to the end of life, remains the same neither in form, nor in structure, nor in the matter of which it is composed : every attribute it possesses is constantly changing, and yet we say that it is always one and the same individual. And if, in this case, we attribute identity without supposing an indivisible immaterial something to underlie and condition that identity, why should we need the sup-position in the case of that succession of changeful phenomena we call the mind?

In fact, we ascribe identity to an individual plant or animal, simply because there has been no moment of time at which we could observe any division of it into parts separated by time or space. Every experience we have of it is as one thing and not as two; and we sum up our experiences in the ascription of identity, although we know quite well that, strictly speaking, it has not been the same for any two moments.

So with the mind. Our perceptions flow in even succession; the impressions of the present moment are inextricably mixed up with the memories of yesterday and the expectations of to-morrow, and all are connected by the links of cause and effect.

".... as the same individual republic may not only change its members, but also its laws and constitutions; in like manner the same person may vary his character and disposition, as well as his impressions and ideas, without losing his identity. Whatever changes he endures, his several parts are still connected by the relation of causation. And, in this view, our identity with regard to the passions serves to corroborate that with regard to the imagination, by the making our distant perceptions influence each other, and by giving us a present concern for our past or future pains or pleasures.

"As memory alone acquaints us with the continuance and extent of this succession of perceptions, 'tis to be considered, upon that account chiefly, as the source of personal identity. Had we no memory we never should have any notion of causation, nor consequently of that chain of causes and effects which constitute our self or person. But having once acquired this notion of causation from the memory, we can extend the same chain of causes, and consequently the identity of our persons, beyond our memory, and can comprehend times, and circumstances, and actions, which we have entirely forgot, but suppose in general to have existed. For how few of our past actions are there of which we have any memory? Who can tell me, for instance, what were his thoughts and actions on the first of January, 1715, the eleventh of March, 1719, and the third of August, 1733? Or will he affirm, because he has entirely forgot the incidents of those days, that the present self is not the same person with the self of that time, and by that means overturn all the most established notions of personal identity? In this view, therefore, memory does not so much *produce* as *discover* personal identity, by showing us the relation of cause and effect among our different perceptions. 'Twill be incumbent on those who affirm that memory produces entirely our personal identity, to give a reason why we can thus extend our identity beyond our memory.

"The whole of this doctrine leads us to a conclusion which

is of great importance in the present affair, viz. that all the nice and subtle questions concerning personal identity can never possibly be decided, and are to be regarded rather as grammatical than as philosophical difficulties. Identity depends on the relations of ideas, and these relations produce identity by means of that easy transition they occasion. But as the relations, and the easiness of the transition may diminish by insensible degrees, we have no just standard by which we can decide any dispute concerning the time when they acquire or lose a title to the name of identity. All the disputes concerning the identity of connected objects are merely verbal, except so far as the relation of parts gives rise to some fiction or imaginary principle of union, as we have already observed.

"What I have said concerning the first origin and uncertainty of our notion of identity, as applied to the human mind, may be extended, with little or no variation, to that of *simplicity*. An object, whose different co-existent parts are bound together by a close relation, operates upon the imagination after much the same manner as one perfectly simple and undivisible, and requires not a much greater stretch of thought in order to its conception. From this similarity of operation we attribute a simplicity to it, and feign a principle of union as the support of this simplicity, and the centre of all the different parts and qualities of the object."—(I. pp. 331–3.)

The final result of Hume's reasoning comes to this: As we use the name of body for the sum of the phenomena which make up our corporeal existence, so we employ the name of soul for the sum of the phenomena which constitute our mental existence; and we have no more reason, in the latter case, than in the former, to suppose that there is anything beyond the phenomena which answers to the name. In the case of the soul, as in that of the body, the idea of substance is a

mere fiction of the imagination, This conclusion
is nothing but a rigorous application of Berkeley's
reasoning concerning matter to mind, and it is
fully adopted by Kant.[1]

Having arrived at the conclusion that the
conception of a soul, as a substantive thing, is
a mere figment of the imagination; and that,
whether it exists or not, we can by no possibility
know anything about it, the inquiry as to the
durability of the soul may seem superfluous.

Nevertheless, there is still a sense in which,
even under these conditions, such an inquiry is
justifiable. Leaving aside the problem of the
substance of the soul, and taking the word "soul"
simply as a name for the series of mental
phenomena which make up an individual mind;
it remains open to us to ask, whether that series
commenced with, or before, the series of
phenomena which constitute the corresponding
individual body; and whether it terminates with
the end of the corporeal series, or goes on after
the existence of the body has ended. And, in
both cases, there arises the further question,
whether the excess of duration of the mental
series over that of the body, is finite or infinite.

---

[1] "Our internal intuition shows no permanent existence, for
the Ego is only the consciousness of my thinking." "There is
no means whatever by which we can learn anything respecting
the constitution of the soul, so far as regards the possibility of
its separate existence."—*Kritik von den Paralogismen der reinen
Vernunft.*

Hume has discussed some of these questions in the remarkable essay " On the Immortality of the Soul," which was not published till after his death, and which seems long to have remained but little known. Nevertheless, indeed, possibly, for that reason, its influence has been manifested in unexpected quarters, and its main arguments have been adduced by archiepiscopal and episcopal authority in evidence of the value of revelation. Dr. Whately,[1] sometime Archbishop of Dublin, paraphrases Hume, though he forgets to cite him ; and Bishop Courtenay's elaborate work,[2] dedicated to the Archbishop, is a development of that prelate's version of Hume's essay.

This little piece occupies only some ten pages, but it is not wonderful that it attracted an acute logician like Whately, for it is a model of clear and vigorous statement. The argument hardly admits of condensation, so that I must let Hume speak for himself :—

" By the mere light of reason it seems difficult to prove the immortality of the soul : the arguments for it are commonly derived either from metaphysical topics, or moral, or physical.

---

[1] *Essays on Some of the Peculiarities of the Christian Religion*, (Essay I. Revelation of a Future State), by Richard Whately, D.D., Archbishop of Dublin.   Fifth Edition, revised, 1846.
[2] *The Future States: their Evidences and Nature ; considered on Principles Physical, Moral, and Scriptural, with the Design of showing the Value of the Gospel Revelation*, by the Right Rev. Reginald Courtenay, D.D., Lord Bishop of Kingston (Jamaica), 1857.

But in reality it is the gospel, and the gospel alone, that has brought *life and immortality* to light.[1]

"1. Metaphysical topics suppose that the soul is immaterial, and that 'tis impossible for thought to belong to a material substance.[2] But just metaphysics teach us that the notion of substance is wholly confused and imperfect; and that we have no other idea of any substance, than as an aggregate of particular qualities inhering in an unknown something. Matter, therefore, and spirit, are at bottom equally unknown, and we cannot determine what qualities inhere in the one or in the other.[3] They likewise teach us that nothing can be decided *à priori* concerning any cause or effect; and that experience, being the only source of our judgments of this nature, we cannot know from any other principle, whether matter, by its structure or arrangement, may not be the cause of thought. Abstract reasonings cannot decide any question of fact or existence. But admitting a spiritual substance to be dispersed throughout the universe, like the ethereal fire of the Stoics, and to be the only inherent subject of thought, we have reason to conclude from *analogy*, that nature uses it after the manner she does the other substance, *matter*. She employs it as a kind of paste or clay; modifies it into a variety of forms

---

[1] "Now that 'Jesus Christ brought life and immortality to light through the Gospel,' and that in the most literal sense, which implies that the revelation of the doctrine is *peculiar* to His Gospel, seems to be at least the most obvious meaning of the Scriptures of the New Testament."—Whately, *l.c.* p. 27.

[2] Compare *Of the Immateriality of the Soul*, Section V. of Part IV., Book I., of the *Treatise*, in which Hume concludes (I. p. 319) that, whether it be material or immaterial, "in both cases the metaphysical arguments for the immortality of the soul are equally inconclusive; and in both cases the moral arguments and those derived from the analogy of nature are equally strong and convincing."

[3] "The question again respecting the materiality of the soul is one which I am at a loss to understand clearly, till it shall have been clearly determined *what matter is*. We know nothing of it, any more than of mind, except its attributes."—Whately, *l.c.* p. 66.

or existences; dissolves after a time each modification, and from its substance erects a new form. As the same material substance may successively compose the bodies of all animals, the same spiritual substance may compose their minds : Their consciousness, or that system of thought which they formed during life, may be continually dissolved by death, and nothing interests them in the new modification. The most positive assertors of the mortality of the soul never denied the immortality of its substance; and that an immaterial substance, as well as a material, may lose its memory or consciousness, appears in part from experience, if the soul be immaterial. Reasoning from the common course of nature, and without supposing any new interposition of the Supreme Cause, which ought always to be excluded from philosophy, *what is incorruptible must also be ingenerable.* The soul, therefore, if immortal, existed before our birth, and if the former existence noways concerned us, neither will the latter. Animals undoubtedly feel, think, love, hate, will, and even reason, though in a more imperfect manner than men : Are their souls also immaterial and immortal?" [1]

Hume next proceeds to consider the moral arguments, and chiefly

". . . those derived from the justice of God, which is supposed to be further interested in the future punishment of the vicious and reward of the virtuous."

But if by the justice of God we mean the same attribute which we call justice in ourselves, then why should either reward or punishment be

---

[1] "None of those who contend for the natural immortality of the soul . . . have been able to extricate themselves from one difficulty, viz. that all their arguments apply, with exactly the same force, to prove an immortality, not only of *brutes*, but even of *plants* ; though in such a conclusion as this they are never willing to acquiesce."—Whately, *l.c.* p. 67.

extended beyond this life ? [1] Our sole means of knowing anything is the reasoning faculty which God has given us; and that reasoning faculty not only denies us any conception of a future state, but fails to furnish a single valid argument in favour of the belief that the mind will endure after the dissolution of the body.

". . . If any purpose of nature be clear, we may affirm that the whole scope and intention of man's creation, so far as we can judge by natural reason, is limited to the present life."

To the argument that the powers of man are so much greater than the needs of this life require, that they suggest a future scene in which they can be employed, Hume replies :—

"If the reason of man gives him great superiority above other animals, his necessities are proportionably multiplied upon him ; his whole time, his whole capacity, activity, courage, and passion, find sufficient employment in fencing against the miseries of his present condition ; and frequently, nay, almost always, are too slender for the business assigned them. A pair of shoes, perhaps, was never yet wrought to the highest degree of perfection that commodity is capable of attaining ; yet it is necessary, at least very useful, that there should be some politicians and moralists, even some geometers, poets and philosophers, among

---

[1] "Nor are we therefore authorised to infer à priori, independent of Revelation, a future state of retribution, from the irregularities prevailing in the present life, since that future state does not account fully for these irregularities. It may explain, indeed, how present evil may be conducive to future good, but not why the good could not be attained without the evil : it may reconcile with our notions of the divine justice the present prosperity of the wicked, but it does not account for the existence of the wicked."—Whately, l.c. pp. 69, 70.

mankind. The powers of men are no more superior to their wants, considered merely in this life, than those of foxes and hares are, compared to *their* wants and to their period of existence. The inference from parity of reason is therefore obvious."

In short, Hume argues that, if the faculties with which we are endowed are unable to discover a future state, and if the most attentive consideration of their nature serves to show that they are adapted to this life and nothing more, it is surely inconsistent with any conception of justice that we should be dealt with as if we had, all along, had a clear knowledge of the fact thus carefully concealed from us. What should we think of the justice of a father, who gave his son every reason to suppose that a trivial fault would only be visited by a box on the ear; and then, years afterwards, put him on the rack for a week for the same fault?

Again, the suggestion arises, if God is the cause of all things, he is responsible for evil as well as for good; and it appears utterly irreconcilable with our notions of justice that he should punish another for that which he has, in fact, done himself. Moreover, just punishment bears a proportion to the offence, while suffering which is infinite is *ipso facto* disproportionate to any finite deed.

"Why then eternal punishment for the temporary offences of so frail a creature as man? Can any one approve of Alex·

ander's rage, who intended to exterminate a whole nation because they had seized his favourite horse Buchephalus ?

" Heaven and hell suppose two distinct species of men, the good and the bad ; but the greatest part of mankind float betwixt vice and virtue. Were one to go round the world with the intention of giving a good supper to the righteous and a sound drubbing to the wicked, he would frequently be embarrassed in his choice, and would find the merits and demerits of most men and women scarcely amount to the value of either." [1]

One can but admire the broad humanity and the insight into the springs of action manifest in this passage. *Comprendre est à moitié pardonner.* The more one knows of the real conditions which determine men's acts the less one finds either to praise or blame. For kindly David Hume, "the damnation of one man is an infinitely greater evil in the universe than the subversion of a thousand million of kingdoms." And he would have felt with his countryman Burns, that even " auld Nickie Ben " should " hae a chance."

As against those who reason for the necessity of a future state, in order that the justice of the Deity may be satisfied, Hume's argumentation appears unanswerable. For if the justice of God

---

[1] "So reason also shows, that for man to expect to earn for himself by the practice of virtue, and claim, as his just right, an immortality of exalted happiness, is a most extravagant and groundless pretension."—Whately, *l.c.* p. 101. On the other hand, however, the Archbishop sees no unreasonableness in a man's earning for himself an immortality of intense unhappiness by the practice of vice. So that life is, naturally, a venture in which you may lose all, but can earn nothing. It may be thought somewhat hard upon mankind if they are pushed into a speculation of this sort, willy-nilly.

resembles what we mean by justice, the bestowal
of infinite happiness for finite well-doing and infinite misery for finite ill-doing, it is in no sense
just. And, if the justice of God does not resemble
what we mean by justice, it is an abuse of
language to employ the name of justice for the
attribute described by it. But, as against those
who choose to argue that there is nothing in what
is known to us of the attributes of the Deity inconsistent with a future state of rewards and
punishments, Hume's pleadings have no force.
Bishop Butler's argument that, inasmuch as the
visitation of our acts by rewards and punishments
takes place in this life, rewards and punishments
must be consistent with the attributes of the
Deity, and therefore may go on as long as the
mind endures, is unanswerable. Whatever exists
is, by the hypothesis, existent by the will of God;
and, therefore, the pains and pleasures which
exist now may go on existing for all eternity,
either increasing, diminishing, or being endlessly
varied in their intensity, as they are now.

It is remarkable that Hume does not refer to
the sentimental arguments for the immortality of
the soul which are so much in vogue at the
present day; and which are based upon our desire
for a longer conscious existence than that which
nature appears to have allotted to us. Perhaps
he did not think them worth notice. For indeed
it is not a little strange, that our strong desire

that a certain occurrence should happen should
be put forward as evidence that it will happen.
If my intense desire to see the friend, from whom
I have parted, does not bring him from the other
side of the world, or take me thither; if the
mother's agonised prayer that her child should
live has not prevented him from dying; experience certainly affords no presumption that the
strong desire to be alive after death, which we
call the aspiration after immortality, is any more
likely to be gratified. As Hume truly says, "All
doctrines are to be suspected which are favoured
by our passions;" and the doctrine, that we are
immortal because we should extremely like to be
so, contains the quintessence of suspiciousness.

In respect of the existence and attributes of
the soul, as of those of the Deity, then, logic
is powerless and reason silent. At the most
we can get no further than the conclusion of
Kant :—

"After we have satisfied ourselves of the vanity of all the
ambitious attempts of reason to fly beyond the bounds of experience, enough remains of practical value to content us. It is
true that no one may boast that he *knows* that God and a future
life exist ; for, if he possesses such knowledge, he is just the
man for whom I have long been seeking. All knowledge
(touching an object of mere reason) can be communicated, and
therefore I might hope to see my own knowledge increased to
this prodigious extent, by his instruction. No ; our conviction
in these matters is not *logical*, but *moral* certainty ; and, inasmuch as it rests upon subjective grounds, (of moral disposition)

I must not even say: *it is* morally certain that there is a God, and so on ; but, *I am* morally certain, and so on. That is to say : the belief in a God and in another world is so interwoven with my moral nature, that the former can no more vanish, than the latter can ever be torn from me.

"The only point to be remarked here is that this act of faith of the intellect (*Vernunftglaube*) assumes the existence of moral dispositions. If we leave them aside, and suppose a mind quite indifferent to moral laws, the inquiry started by reason becomes merely a subject for speculation ; and [the conclusion attained] may then indeed be supported by strong arguments from analogy, but not by such as are competent to overcome persistent scepticism.

"There is no one, however, who can fail to be interested in these questions. For, although he may be excluded from moral influences by the want of a good disposition, yet, even in this case, enough remains to lead him to fear a divine existence and a future state. To this end, no more is necessary than that he can at least have no certainty that there is no such being, and no future life ; for, to make this conclusion demonstratively certain, he must be able to prove the impossibility of both ; and this assuredly no rational man can undertake to do. This negative belief, indeed, cannot produce either morality or good dispositions, but can operate in an analogous fashion, by powerfully repressing the outbreak of evil tendencies.

"But it will be said, is this all that Pure Reason can do when it gazes out beyond the bounds of experience ? Nothing more than two articles of faith ? Common sense could achieve as much without calling the philosophers to its counsels !

"I will not here speak of the service which philosophy has rendered to human reason by the laborious efforts of its criticism, granting that the outcome proves to be merely negative : about that matter something is to be said in the following section. But do you then ask, that the knowledge which interests all men shall transcend the common understanding and be discovered for you only by philosophers ? The very thing which you make a reproach, is the best confirmation of the justice of the previous conclusions, since it shows that which could not, at first, have been anticipated; namely, that in those matters which concern all men alike, nature is not guilty of distributing her gifts with partiality; and that the highest philosophy, in dealing with the most important concerns of humanity, is able to take us no further than the guidance which she affords to the commonest understanding."[1]

In short, nothing can be proved or disproved respecting either the distinct existence, the substance, or the durability of the soul. So far, Kant is at one with Hume. But Kant adds, as you cannot disprove the immortality of the soul, and as the belief therein is very useful for moral purposes, you may assume it. To which, had Hume lived half a century later, he would probably have replied, that, if morality has no better foundation than an assumption, it is not likely to bear much strain; and, if it has a better foundation, the assumption rather weakens than strengthens it.

As has been already said, Hume is not content with denying that we know anything about the existence or the nature of the soul; but he carries the war into the enemy's camp, and accuses those who affirm the immateriality, simplicity, and indivisibility of the thinking substance of atheism and Spinozism, which are assumed to be convertible terms.

The method of attack is ingenious. Observation appears to acquaint us with two different systems of beings, and both Spinoza and orthodox

[1] *Kritik der reinen Vernunft.* Ed. Hartenstein, p. 547

philosophers agree, that the necessary substratum of each of these is a substance, in which the phenomena adhere, or of which they are attributes or modes.

" I observe first the universe of objects or of body ; the sun, moon, and stars ; the earth, seas, plants, animals, men, ships, houses, and other productions either of art or of nature. Here Spinoza appears, and tells me that these are only modifications and that the subject in which they inhere is simple, uncompounded, and indivisible. After this I consider the other system of beings, viz. the universe of thought, or my impressions and ideas. Then I observe another sun, moon, and stars ; an earth and seas, covered and inhabited by plants and animals, towns, houses, mountains, rivers ; and, in short, everything I can discover or conceive in the first system. Upon my inquiring concerning these, theologians present themselves, and tell me that these also are modifications, and modifications of one simple, uncompounded, and indivisible substance. Immediately upon which I am deafened with the noise of a hundred voices, that treat the first hypothesis with detestation and scorn, and the second with applause and veneration. I turn my attention to these hypotheses to see what may be the reason of so great a partiality ; and find that they have the same fault of being unintelligible, and that, as far as we can understand them, they are so much alike, that 'tis impossible to discover any absurdity in one, which is not common to both of them." —(I. p. 309.)

For the manner in which Hume makes his case good, I must refer to the original. Plain people may rest satisfied that both hypotheses are unintelligible, without plunging any further among syllogisms, the premisses of which convey no meaning, while the conclusions carry no conviction.

CHAPTER X

IN the opening paragraphs of the third part of the second book of the "Treatise," Hume gives a description of the will.

"Of all the immediate effects of pain and pleasure there is none more remarkable than the *will;* and though, properly speaking, it be not comprehended among the passions, yet as the full understanding of its nature and properties is necessary to the explanation of them, we shall here make it the subject of our inquiry. I desire it may be observed, that, by the *will,* I mean nothing but *the internal impression we feel, and are conscious of, when we knowingly give rise to any new motion of our body, or new perception of our mind.* This impression, like the preceding ones of pride and humility, love and hatred, 'tis impossible to define, and needless to describe any further."—(II. p. 150.)

This description of volition may be criticised on various grounds. More especially does it seem defective in restricting the term "will" to that feeling which arises when we act, or appear to act, as causes: for one may will to strike, with-

out striking; or to think of something which we
have forgotten.

Every volition is a complex idea composed of
two elements: the one is the idea of an action;
the other is a desire for the occurrence of that
action. If I will to strike, I have an idea of a
certain movement, and a desire that that movement should take place; if I will to think of any
subject, or, in other words, to attend to that subject, I have an idea of the subject and a strong
desire that it should remain present to my consciousness. And so far as I can discover, this
combination of an idea of an object with an
emotion, is everything that can be directly
observed in an act of volition. So that Hume's
definition may be amended thus: Volition is the
impression which arises when the idea of a bodily
or mental action is accompanied by the desire that
the action should be accomplished. It differs
from other desires simply in the fact, that we
regard ourselves as possible causes of the action
desired.

Two questions arise, in connexion with the
observation of the phenomenon of volition, as
they arise out of the contemplation of all other
natural phenomena. Firstly, has it a cause;
and, if so, what is its cause? Secondly, is it
followed by any effect, and if so, what effect does
it produce?

Hume points out, that the nature of the phenomena we consider can have nothing to do with the origin of the conception that they are connected by the relation of cause and effect. For that relation is nothing but an order of succession, which, so far as our experience goes, is invariable; and it is obvious that the nature of phenomena has nothing to do with their order. Whatever it is that leads us to seek for a cause for every event, in the case of the phenomena of the external world, compels us, with equal cogency, to seek it in that of the mind.

The only meaning of the law of causation, in the physical world, is, that it generalises universal experience of the order of that world; and, if experience shows a similar order to obtain among states of consciousness, the law of causation will properly express that order.

That such an order exists, however, is acknowledged by every sane man:

"Our idea, therefore, of necessity and causation, arises entirely from the uniformity observable in the operations of nature, where similar objects are constantly conjoined together, and the mind is determined by custom to infer the one from the appearance of the other. These two circumstances form the whole of that necessity which we ascribe to matter. Beyond the constant *conjunction* of similar objects and the consequent *inference* from one to the other, we have no notion of any necessity of connexion.

"If it appear, therefore, what all mankind have ever allowed, without any doubt or hesitation, that these two circumstances take place in the voluntary actions of men, and in the operations of mind, it must follow that all mankind have

ever agreed in the doctrine of necessity, and that they have
hitherto disputed merely from not understanding each other."
—(IV. p. 97.)

But is this constant conjunction observable in
human actions ? A student of history could give
but one answer to this question :

"Ambition, avarice, self-love, vanity, friendship, generosity,
public spirit : these passions, mixed in various degrees, and
distributed through society, have been, from the beginning of
the world, and still are, the source of all the actions and enter-
prizes which have ever been observed among mankind. Would
you know the sentiments, inclinations, and course of life of the
Greeks and Romans ? Study well the temper and actions of the
French and English. You cannot be much mistaken in trans-
ferring to the former *most* of the observations which you have
made with regard to the latter. Mankind are so much the
same, in all times and places, that history informs us of nothing
new or strange in this particular. Its chief use is only to discover the constant and universal principles of human nature,
by showing men in all varieties of circumstances and situations,
and furnishing us with materials from which we may form our
observations, and become acquainted with the regular springs of
human action and behaviour. These records of wars, intrigues,
factions, and revolutions are so many collections of experiments,
by which the politician or moral philosopher fixes the principles
of his science, in the same manner as the physician or natural
philosopher becomes acquainted with the nature of plants,
minerals, and other external objects, by the experiments which
he forms concerning them. Nor are the earth, air, water, and
other elements examined by Aristotle and Hippocrates more like
to those which at present lie under our observation, than the
men described by Polybius and Tacitus are to those who now
govern the world."—(IV. pp. 97-8.)

Hume proceeds to point out that the value set
upon experience in the conduct of affairs, whether

of business or of politics, involves the acknowledgment that we base our expectation of what men will do, upon our observation of what they have done; and, that we are as firmly convinced of the fixed order of thoughts as we are of that of things. And, if it be urged that human actions not unfrequently appear unaccountable and capricious, his reply is prompt:—

"I grant it possible to find some actions which seem to have no regular connexion with any known motives, and are exceptions to all the measures of conduct which have ever been established for the government of men. But if one could willingly know what judgment should be formed of such irregular and extraordinary actions, we may consider the sentiments commonly entertained with regard to those irregular events which appear in the course of nature, and the operations of external objects. All courses are not conjoined to their usual effects with like uniformity. An artificer, who handles only dead matter, may be disappointed in his aim, as well as the politician who directs the conduct of sensible and intelligent agents.

"The vulgar, who take things according to their first appearance, attribute the uncertainty of events to such an uncertainty in the causes as make the latter often fail of their usual influence, though they meet with no impediment to their operation. But philosophers, observing that, almost in every part of nature, there is contained a vast variety of springs and principles, which are hid, by reason of their minuteness or remoteness, find that it is at least possible the contrariety of events may not proceed from any contingency in the cause, but from the secret operation of contrary causes. This possibility is converted into certainty by further observation, when they remark that, upon an exact scrutiny, a contrariety of effects always betrays a contrariety of causes, and proceeds from their mutual opposition. A peasant can give no better reason for

the stopping of any clock or watch, than to say that it does not
commonly go right. But an artist easily perceives that the same
force in the spring or pendulum has always the same influence
on the wheels ; but fails of its usual effect, perhaps by reason
of a grain of dust, which puts a stop to the whole movement.
From the observation of several parallel instances, philosophers
form a maxim, that the connexion between all causes and
effects is equally necessary, and that its seeming uncertainty in
some instances proceeds from the secret opposition of contrary
causes."—(IV. pp. 101-2.)

So with regard to human actions :—

" The internal principles and motives may operate in a uniform manner, notwithstanding these seeming irregularities ; in
the same manner as the winds, rains, clouds, and other variations of the weather are supposed to be governed by steady
principles ; though not easily discoverable by human sagacity
and inquiry."—(IV. p. 103.)

Meteorology, as a science, was not in existence
in Hume's time, or he would have left out the
"supposed to be." In practice, again, what difference does any one make between natural and
moral evidence ?

" A prisoner who has neither money nor interest, discovers
the impossibility of his escape, as well, when he considers the
obstinacy of the goaler, as the walls and bars with which he is
surrounded ; and, in all attempts for his freedom, chooses
rather to work upon the stone and iron of the one, than upon
the inflexible nature of the other. The same prisoner, when
conducted to the scaffold, foresees his death as certainly from
the constancy and fidelity of his guards, as from the operation
of the axe or wheel. His mind runs along a certain train of
ideas : The refusal of the soldiers to consent to his escape ; the
action of the executioner ; the separation of the head and body ;

bleeding, convulsive motions, and death. Here is a connected
chain of natural causes and voluntary actions; but the mind
feels no difference between them, in passing from one link
to another, nor is less certain of the future event, than if it
were connected with the objects presented to the memory or
senses, by a train of causes cemented together by what we are
pleased to call a *physical* necessity. The same experienced
union has the same effect on the mind, whether the united
objects be motives, volition, and actions; or figure and motion.
We may change the names of things, but their nature and
their operation on the understanding never change."—(IV. pp.
105-6.)

But, if the necessary connexion of our acts
with our ideas has always been acknowledged in
practice, why the proclivity of mankind to deny it
words?

" If we examine the operations of body, and the production
of effects from their causes, we shall find that all our faculties
can never carry us further in our knowledge of this relation,
than barely to observe, that particular objects are *constantly
conjoined* together, and that the mind is carried, by a *customary
transition*, from the appearance of the one to the belief of the
other. But though this conclusion concerning human ignorance be the result of the strictest scrutiny of this subject, men
still entertain a strong propensity to believe, that they penetrate
further into the province of nature, and perceive something
like a necessary connexion between cause and effect. When,
again, they turn their reflections towards the operations of their
own minds, and *feel* no such connexion between the motive and
the action; they are thence apt to suppose, that there is a
difference between the effects which result from material force,
and those which arise from thought and intelligence. But,
being once convinced, that we know nothing of causation of any
kind, than merely the *constant conjunction* of objects, and the
consequent *inference* of the mind from one to another, and finding that these two circumstances are universally allowed to have

place in voluntary actions; we may be more easily led to own
the same necessity common to all causes."—(IV. pp. 107, 8.)

The last asylum of the hard-pressed advocate of
the doctrine of uncaused volition is usually, that,
argue as you like, he has a profound and ineradicable consciousness of what he calls the freedom of
his will. But Hume follows him even here,
though only in a note, as if he thought the ex-
tinction of so transparent a sophism hardly worthy
of the dignity of his text.

"The prevalence of the doctrine of liberty may be accounted
for from another cause, viz. a false sensation, or seeming experience, which we have, or may have, of liberty or indifference in
many of our actions. The necessity of any action, whether of
matter, or of mind, is not, properly speaking, a quality in the
agent, but in any thinking or intelligent being who may consider the action ; and it consists chiefly in the determination of
his thoughts to infer the existence of that action from some
preceding objects ; as liberty, when opposed to necessity, is
nothing but the want of that determination, and a certain looseness or indifference which we feel in passing, or not passing,
from the idea of any object to the idea of any succeeding one.
Now we may observe that though, in *reflecting* on human
actions, we seldom feel such looseness or indifference, but are
commonly able to infer them with considerable certainty from
their motives, and from the dispositions of the agent ; yet it
frequently happens that in *performing* the actions themselves,
we are sensible of something like it : And as all resembling
objects are taken for each other, this has been employed as
demonstrative and even intuitive proof of human liberty. We
feel that our actions are subject to our will on most occasions ;
and imagine we feel, that the will itself is subject to nothing,
because, when by a denial of it we are provoked to try, we feel
that it moves easily every way, and produces an image of itself

(or a *Velleity* as it is called in the schools), even on that side on which it did not settle. This image or faint notion, we persuade ourselves, could at that time have been completed into the thing itself; because, should that be denied, we find upon a second trial that at present it can. We consider not that the fantastical desire of showing liberty is here the motive of our actions."—(IV. p. 110, *note.*)

Moreover the moment the attempt is made to give a definite meaning to the words, the supposed opposition between free will and necessity turns out to be a mere verbal dispute.

"For what is meant by liberty, when applied to voluntary actions? We cannot surely mean, that actions have so little connexion with motive, inclinations, and circumstances, that one does not follow with a certain degree of uniformity from the other, and that one affords no inference by which we can conclude the existence of the other. For these are plain and acknowledged matters of fact. By liberty, then, we can only mean *a power of acting or not acting according to the determinations of the will;* that is, if we choose to remain at rest, we may ; if we choose to move, we also may   Now this hypothetical liberty is universally allowed to belong to every one who is not a prisoner and in chains. Here then is no subject of dispute."—(IV. p. 111.)

Half the controversies about the freedom of the will would have had no existence, if this pithy paragraph had been well pondered by those who oppose the doctrine of necessity. For they rest upon the absurd presumption that the proposition, " I can do as I like," is contradictory to the doctrine of necessity. The answer is ; nobody doubts that, at any rate within certain limits, you can do as

you like.   But what determines your likings and
dislikings ?   Did you make your own constitution ?
Is it your contrivance that one thing is pleasant
and another is painful ?   And even if it were, why
did you prefer to make it after the one fashion
rather than the other ?   The passionate assertion
of the consciousness of their freedom, which is the
favourite refuge of the opponents of the doctrine
of necessity, is mere futility, for nobody denies it.
What they really have to do, if they would upset the necessarian argument, is to prove that
they are free to associate any emotion whatever
with any idea whatever; to like pain as much as
pleasure; vice as much as virtue; in short, to
prove, that, whatever may be the fixity of order of
the universe of things, that of thought is given
over to chance.

In the second part of this remarkable essay,
Hume considers the real, or supposed, immoral consequences of the doctrine of necessity, premising
the weighty observation that

"When any opinion leads to absurdity, it is certainly false ;
but it is not certain that an opinion is false because it is of
dangerous consequence."—(IV. p. 112.)

And, therefore, that the attempt to refute an
opinion by a picture of its dangerous consequences
to religion and morality, is as illogical as it is
reprehensible.

It is said, in the first place, that necessity destroys responsibility; that, as it is usually put, we have no right to praise or blame actions that cannot be helped. Hume's reply amounts to this, that the very idea of responsibility implies the belief in the necessary connexion of certain actions with certain states of the mind. A person is held responsible only for those acts which are preceded by a certain intention; and, as we cannot see, or hear, or feel, an intention, we can only reason out its existence on the principle that like effects have like causes.

If a man is found by the police busy with "jemmy" and dark lantern at a jeweller's shop door over night, the magistrate before whom he is brought the next morning, reasons from those effects to their causes in the fellow's burglarious ideas and volitions, with perfect confidence, and punishes him accordingly. And it is quite clear that such a proceeding would be grossly unjust, if the links of the logical process were other than necessarily connected together. The advocate who should attempt to get the man off on the plea that his client need not necessarily have had a felonious intent, would hardly waste his time more, if he tried to prove that the sum of all the angles of a triangle is not two right angles, but three.

A man's moral responsibility for his acts has, in fact, nothing to do with the causation of these acts, but depends on the frame of mind which

accompanies them.   Common language tells us
this, when it uses "well disposed" as the equivalent of "good," and "evil-minded" as that of
" wicked."   If A does something which puts B in
a violent passion, it is quite possible to admit that
B's passion is the necessary consequence of A's
act, and yet to believe that B's fury is morally
wrong, or that he ought to control it.   In fact, a
calm bystander would reason with both on the
assumption of moral necessity.   He would say to
A, " You were wrong in doing a thing which you
knew (that is, of the necessity of which you were
convinced) would irritate B."   And he would say
to B, " You are wrong to give way to passion, for
you know its evil effects "—that is the necessary
connection between yielding to passion and evil.

So far, therefore, from necessity destroying
moral responsibility, it is the foundation of all
praise and blame ; and moral admiration reaches
its climax in the ascription of necessary goodness
to the Deity.

To the statement of another consequence of the
necessarian doctrine, that, if there be a God, he
must be the cause of all evil as well as of all good,
Hume gives no real reply—probably because none
is possible.   But then, if this conclusion is distinctly and unquestionably deducible from the
doctrine of necessity, it is no less unquestionably
a direct consequence of every known form of
monotheism.   If God is the cause of all things,

he must be the cause of evil among the rest; if
he is omniscient, he must have the fore-knowledge
of evil; if he is almighty, he must possess the
power of preventing, or of extinguishing evil.
And to say that an all-knowing and all-powerful
being is not responsible for what happens, because
he only permits it, is, under its intellectual aspect,
a piece of childish sophistry; while, as to the
moral look of it, one has only to ask any decently
honourable man, whether, under like circumstances, he would try to get rid of his responsibility
by such a plea.

Hume's "Inquiry" appeared in 1748. He does
not refer to Anthony Collins' essay on Liberty,
published thirty-three years before, in which the
same question is treated to the same effect, with
singular force and lucidity. It may be said,
perhaps, that it is not wonderful that the two
freethinkers should follow the same line of reasoning; but no such theory will account for the fact
that in 1754, the famous Calvinistic divine,
Jonathan Edwards, President of the College of
New Jersey, produced, in the interests of the
straitest orthodoxy, a demonstration of the necessarian thesis, which has never been equalled in
power, and certainly has never been refuted.

In the ninth section of the fourth part of
Edwards's "Inquiry," he has to deal with the
Arminian objection to the Calvinistic doctrine
that "it makes God the author of sin"; and it is

curious to watch the struggle between the theological controversialist, striving to ward off an
admission which he knows will be employed to
damage his side, and the acute logician, conscious
that, in some shape or other, the admission must
be made. Beginning with a *tu quoque*, that the
Arminian doctrine involves consequences as bad
as the Calvinistic view, he proceeds to object to
the term "author of sin," though he ends by
admitting that, in a certain sense, it is applicable;
he proves from Scripture, that God is the disposer
and orderer of sin ; and then, by an elaborate false
analogy with the darkness resulting from the
absence of the sun, endeavours to suggest that he
is only the author of it in a negative sense; and,
finally, he takes refuge in the conclusion that,
though God is the orderer and disposer of those
deeds which, considered in relation to their agents,
are morally evil, yet inasmuch as His purpose has
all along been infinitely good, they are not evil
relatively to Him.

And this, of course, may be perfectly true ; but
if true, it is inconsistent with the attribute of
Omnipotence. It is conceivable that there should
be no evil in the world ; that which is conceivable
is certainly possible ; if it were possible for evil to
be non-existent, the maker of the world, who,
though foreknowing the existence of evil in that
world, did not prevent it, either did not really
desire it should not exist, or could not prevent its

existence. It might be well for those who inveigh against the logical consequences of necessarianism to bethink them of the logical consequences of theism; which are not only the same, when the attribute of Omniscience is ascribed to the Deity, but which bring out, from the existence of moral evil, a hopeless conflict between the attributes of Infinite Benevolence and Infinite Power, which, with no less assurance, are affirmed to appertain to the Divine Being.

Kant's mode of dealing with the doctrine of necessity is very singular. That the phenomena of the mind follow fixed relations of cause and effect is, to him, as unquestionable as it is to Hume. But then there is the *Ding an sich*, the *Noumenon*, or Kantian equivalent for the substance of the soul. This, being out of the phenomenal world, is subject to none of the laws of phenomena, and is consequently as absolutely free, and as completely powerless, as a mathematical point, *in vacuo*, would be. Hence volition is uncaused, so far as it belongs to the noumenon; but, necessary, so far as it takes effect in the phenomenal world.

Since Kant is never weary of telling us that we know nothing whatever, and can know nothing, about the noumenon, except as the hypothetical subject of any number of negative predicates; the information that it is free, in the sense of being out of reach of the law of causation, is about as

valuable as the assertion that it is neither gray,
nor blue, nor square. For practical purposes, it
must be admitted that the inward possession of
such a noumenal libertine does not amount to
much for people whose actual existence is made
up of nothing but definitely regulated phenomena.
When the good and evil angels fought for the
dead body of Moses, its presence must have been
of about the same value to either of the contend-
ing parties, as that of Kant's noumenon, in the
battle of impulses which rages in the breast of
man. Metaphysicians, as a rule, are sadly deficient
in the sense of humour; or they would surely
abstain from advancing propositions which, when
stripped of the verbiage in which they are disguised, appear to the profane eye to be bare
shams, naked but not ashamed.

CHAPTER XI

In his autobiography, Hume writes :—

> "In the same year [1752] was published at London my
> 'Inquiry Concerning the Principles of Morals'; which in my
> own opinion (who ought not to judge on that subject) is of all my
> writings, historical, philosophical, and literary, incomparably
> the best. It came unnoticed and unobserved into the world."

It may commonly be noticed that the relative
value which an author ascribes to his own works
rarely agrees with the estimate formed of them
by his readers; who criticise the products, without either the power, or the wish, to take into
account the pains which they may have cost the
producer. Moreover, the clear and dispassionate
common sense of the "Inquiry Concerning the
Principles of Morals" may have tasted flat after
the highly-seasoned "Inquiry Concerning the
Human Understanding." Whether the public
like to be deceived, or not, may be open to question; but it is beyond a doubt that they love to

be shocked in a pleasant and mannerly way. Now Hume's speculations on moral questions are not so remote from those of respectable professors, like Hutcheson, or saintly prelates, such as Butler, as to present any striking novelty. And they support the cause of righteousness in a cool, reasonable, indeed slightly patronising fashion, eminently in harmony with the mind of the eighteenth century; which admired virtue very much, if she would only avoid the rigour which the age called fanaticism, and the fervour which it called enthusiasm.

Having applied the ordinary methods of scientific inquiry to the intellectual phenomena of the mind, it was natural that Hume should extend the same mode of investigation to its moral phenomena; and, in the true spirit of a natural philosopher, he commences by selecting a group of those states of consciousness with which every one's personal experience must have made him familiar: in the expectation that the discovery of the sources of moral approbation and disapprobation, in this comparatively easy case, may furnish the means of detecting them when they are more recondite.

" We shall analyse that complication of mental qualities which form what, in common life, we call PERSONAL MERIT : We shall consider every attribute of the mind, which renders a man an object either of esteem and affection, or of hatred and contempt ; every habit or sentiment or faculty, which if ascribed to any person, implies either praise or blame, and may enter into any panegyric or satire of his character and manners.   The quick sensibility, which, on this head, is so universal among mankind, gives a philosopher sufficient assurance that he can never be considerably mistaken in framing the catalogue, or incurs any danger of misplacing the objects of his contemplation : He needs only enter into his own breast for a moment, and consider whether he should or should not desire to have this or that quality assigned to him, and whether such or such an imputation would proceed from a friend or an enemy. The very nature of language guides us almost infallibly in forming a judgment of this nature ; and as every tongue possesses one set of words which are taken in a good sense, and another in the opposite, the least acquaintance with the idiom suffices, without any reasoning, to direct us in collecting and arranging the estimable or blamable qualities of men. The only object of reasoning is to discover the circumstances on both sides, which are common to these qualities ; to observe that particular in which the estimable qualities agree on the one hand, and the blamable on the other, and thence to reach the foundation of ethics, and find their universal principles, from which all censure or approbation is ultimately derived. As this is a question of fact, not of abstract science, we can only expect success by following the experimental method, and deducing general maxims from a comparison of particular instances. The other scientifical method, where a general abstract principle is first established, and is afterwards branched out into a variety of inferences and conclusions, may be more perfect in itself, but suits less the imperfection of human nature, and is a common source of illusion and mistake, in this as well as in other subjects. Men are now cured of their passion for hypotheses and systems in natural philosophy, and will hearken to no arguments but those which are derived from experience. It is full time they should attempt a like reformation in all moral disquisitions ; and reject every system of ethics, however subtile or ingenious, which is not founded on fact and observation."—(IV. pp. 242—4.)

No qualities give a man a greater claim to personal merit than benevolence and justice; but if we inquire why benevolence deserves so much praise, the answer will certainly contain a large reference to the utility of that virtue to society; and as for justice, the very existence of the virtue implies that of society; public utility is its sole origin; and the measure of its usefulness is also the standard of its merit. If every man possessed everything he wanted, and no one had the power to interfere with such possession; or if no man desired that which could damage his fellow-man, justice would have no part to play in the universe. But as Hume observes:—

"In the present disposition of the human heart, it would perhaps be difficult to find complete instances of such enlarged affections; but still we may observe that the case of families approaches towards it; and the stronger the mutual benevolence is among the individuals, the nearer it approaches, till all distinction of property be in a great measure lost and confounded among them. Between married persons, the cement of friendship is by the laws supposed so strong, as to abolish all division of possessions, and has often, in reality, the force assigned to it.[1] And it is observable that, during the ardour of new enthusiasms, when every principle is inflamed into extravagance, the community of goods has frequently been attempted; and nothing but experience of its inconveniences, from the returning or disguised selfishness of men, could make the imprudent fanatics adopt anew the ideas of justice and separate property. So true is it that this virtue derives its existence entirely from its necessary *use* to the intercourse and social state of mankind."[1]— (IV. p. 256.)

"Were the human species so framed by nature as that each

---

[1] Family affection in the eighteenth century may have been stronger than in the nineteenth; but Hume's bachelor inexperience can surely alone explain his strange account of the suppositions of the marriage law of that day, and their effects. The law certainly abolished all division of possessions, but it did so by making the husband sole proprietor.

individual possessed within himself every faculty requisite both for his own preservation and for the propagation of his kind : Were all society and intercourse cut off between man and man by the primary intention of the Supreme Creator : It seems evident that so solitary a being would be as much incapable of justice as of social discourse and conversation. Where mutual regard and forbearance serve to no manner of purpose, they would never direct the conduct of any reasonable man. The headlong course of the passions would be checked by no reflection on future consequenqes. And as each man is here supposed to love himself alone, and to depend only on himself and his own activity for safety and happiness, he would, on every occasion, to the utmost of his power, challenge the preference above every other being, to none of which he is bound by any ties, either of nature or of interest.

"But suppose the conjunction of the sexes to be established in nature, a family immediately arises ; and particular rules being found requisite for its subsistence, these are immediately embraced, though without comprehending the rest of mankind within their prescriptions. Suppose that several families unite together in one society, which is totally disjoined from all others, the rules which preserve peace and order enlarge themselves to the utmost extent of that society ; but becoming then entirely useless, lose their force when carried one step further. But again, suppose that several distinct societies maintain a kind of intercourse for mutual convenience and advantage, the boundaries of justice still grow larger, in proportion to the largeness of men's views and the force of their mutual connexion. History, experience, reason, sufficiently instruct us in this natural progress of human sentiments, and in the gradual enlargement of our regard to justice in proportion as we become acquainted with the extensive utility of that virtue."—(IV. pp. 262—4.)

The moral obligation of justice and the rights of property are by no means diminished by this exposure of the purely utilitarian basis on which they rest :—

"For what stronger foundation can be desired or conceived

for any duty, than to observe that human society, or even
human nature, could not subsist without the establishment of
it, and will still arrive at greater degrees of happiness and
perfection, the more inviolable the regard is which is paid to
that duty ?

"The dilemma seems obvious : As justice evidently tends
to promote public utility, and to support civil society, the
sentiment of justice is either derived from our reflecting on
that tendency, or, like hunger, thirst, and other appetites, resentment, love of life, attachment to offspring, and other
passions, arises from a simple original instinct in the human
heart, which nature has implanted for like salutary purposes.
If the latter be the case, it follows that property, which is the
object of justice, is also distinguished by a simple original
instinct, and is not ascertained by any argument or reflection.
But who is there that ever heard of such an instinct? Or is
this a subject in which new discoveries can be made? We may
as well expect to discover in the body new senses which had
before escaped the observation of all mankind."—(IV. pp. 273–
4.)

The restriction of the object of justice to property, in this passage, is singular. Pleasure and
pain can hardly be included under the term property, and yet justice surely deals largely with the
withholding of the former, or the infliction of the
latter, by men on one another. If a man bars
another from a pleasure which he would otherwise
enjoy, or actively hurts him without good reason,
the latter is said to be injured as much as if his
property had been interfered with. Here, indeed,
it may be readily shown, that it is as much the
interest of society that men should not interfere
with one another's freedom, or mutually inflict
positive or negative pain, as that they should not

meddle with one another's property ; and hence
the obligation of justice in such matters may be
deduced. But, if a man merely thinks ill of
another, or feels maliciously towards him without
due cause, he is properly said to be unjust. In
this case it would be hard to prove that any injury
is done to society by the evil thought ; but there
is no question that it will be stigmatised as an
injustice ; and the offender himself, in another
frame of mind, is often ready enough to admit
that he has failed to be just towards his neighbour.
However, it may plausibly be said, that so slight a
barrier lies between thought and speech, that any
moral quality attached to the latter is easily
transferred to the former ; and that, since open
slander is obviously opposed to the interests of
society, injustice of thought, which is silent
slander, must become inextricably associated with
the same blame.

But, granting the utility to society of all kinds
of benevolence and justice, why should the
quality of those virtues involve the sense of moral
obligation ?

Hume answers this question in the fifth section
entitled, " Why Utility Pleases." He repudiates
the deduction of moral approbation from self-love,
and utterly denies that we approve of benevolent
or just actions because we think of the benefits
which they are likely to confer indirectly on ourselves. The source of the approbation with which
we view an act useful to society must be sought

elsewhere; and, in fact, is to be found in that feeling which is called sympathy.

"No man is absolutely indifferent to the happiness and misery of others. The first has a natural tendency to give pleasure, the second pain. This every one may find in himself. It is not probable that these principles can be resolved into principles more simple and universal, whatever attempts may have been made for that purpose."—(IV. p. 294, *Note.*)

Other men's joys and sorrows are not spectacles at which we remain unmoved :—

". . . The view of the former, whether in its causes or effects, like sunshine, or the prospect of well-cultivated plains (to carry our pretensions no higher) communicates a secret joy and satisfaction ; the appearance of the latter, like a lowering cloud or barren landscape, throws a melancholy damp over the imagination. And this concession being once made, the difficulty is over ; and a natural unforced interpretation of the phenomena of human life will afterwards, we hope, prevail among all speculative inquirers."—(IV. p. 320.)

The moral approbation, therefore, with which we regard acts of justice or benevolence rests upon their utility to society, because the perception of that utility or, in other words, of the pleasure which they give to other men, arouses a feeling of sympathetic pleasure in ourselves. The feeling of obligation to be just, or of the duty of justice, arises out of that association of moral approbation or disapprobation with one's own actions, which is what we call conscience. To fail in justice, or in benevolence, is to be displeased with one's self. But happiness is impossible without inward self-

approval; and, hence, every man who has any
regard to his own happiness and welfare, will find
his best reward in the practice of every moral
duty. On this topic Hume expends much eloquence.

"But what philosophical truths can be more advantageous
to society than these here delivered, which represent virtue in
all her genuine and most engaging charms, and make us
approach her with ease, familiarity, and affection? The
dismal dress falls off, with which many divines and some
philosophers have covered her; and nothing appears but gentle-
ness, humanity, beneficence, affability; nay, even at proper
intervals, play, frolic, and gaiety. She talks not of useless
austerities and rigours, suffering and self-denial. She declares
that her sole purpose is to make her votaries, and all mankind,
during every period of their existence, if possible, cheerful,
and happy; nor does she ever willingly part with any pleasure
but in hopes of ample compensation in some other period of
their lives. The sole trouble which she demands is that of
just calculation, and a steady preference of the greater
happiness. And if any austere pretenders approach her,
enemies to joy and pleasure, she either rejects them as
hypocrites and deceivers, or if she admit them in her train,
they are ranked, however, among the least favoured of her
votaries.

"And, indeed, to drop all figurative expression, what hopes
can we ever have of engaging mankind to a practice which
we confess full of austerity and rigour? Or what theory of
morals can ever serve any useful purpose, unless it can show, by
a particular detail, that all the duties which it recommends are
also the true interest of each individual? The peculiar advantage of the foregoing system seems to be, that it furnishes
proper mediums for that purpose."—(IV. p. 360.)

In this pæan to virtue, there is more of the
dance measure than will sound appropriate in the

ears of most of the pilgrims who toil painfully, not without many a stumble and many a bruise, along the rough and steep roads which lead to the higher life.

Virtue is undoubtedly beneficent; but the man is to be envied to whom her ways seem in anywise playful. And though she may not talk much about suffering and self-denial, her silence on that topic may be accounted for on the principle *ça va sans dire*. The calculation of the greatest happiness is not performed quite so easily as a rule or three sum; while, in the hour of temptation, the question will crop up, whether, as something has to be sacrificed, a bird in the hand is not worth two in the bush; whether it may not be as well to give up the problematical greater happiness in the future, for a certain great happiness in the present, and

> " Buy the merry madness of one hour
> With the long irksomeness of following time." [1]

If mankind cannot be engaged in practices " full of austerity and rigour," by the love of righteousness and the fear of evil, without seeking for other compensation than that which flows from the gratification of such love and the consciousness of escape from debasement, they are in a bad case. For they will assuredly find that virtue presents no very close likeness to the sportive leader of the joyous hours in Hume's rosy picture; but that she

[1] Ben Jonson's *Cynthia's Revels*, act i.

is an awful Goddess, whose ministers are the
Furies, and whose highest reward is peace.

It is not improbable that Hume would have
qualified all this as enthusiasm or fanaticism, or
both ; but he virtually admits it :—

"Now, as virtue is an end, and is desirable on its own
account, without fee or reward, merely for the immediate satisfaction which it conveys, it is requisite that there should be some
sentiment which it touches ; some internal taste or feeling, or
whatever you please to call it, which distinguishes moral good
and evil, and which embraces the one and rejects the other.

"Thus the distinct boundaries and offices of *reason* and of
taste are easily ascertained. The former conveys the knowledge
of truth and falsehood : The latter gives the sentiment of
beauty and deformity, vice and virtue. The one discovers objects
as they really stand in nature, without addition or diminution :
The other has a productive faculty, and gilding and staining all
natural objects with the colours borrowed from internal sentiment, raises in a manner a new creation. Reason being cool
and disengaged, is no motive to action, and directs only the
impulse received from appetite or inclination, by showing us the
means of attaining happiness or avoiding misery. Taste, as it
gives pleasure or pain, and thereby constitutes happiness or
misery, becomes a motive to action, and is the first spring
or impulse to desire and volition. From circumstances
and relations known or supposed, the former leads us to
the discovery of the concealed and unknown. After all
circumstances and relations are laid before us, the latter
makes us feel from the whole a new sentiment of blame or
approbation. The standard of the one, being founded on the
nature of things, is external and inflexible, even by the will of
the Supreme Being : The standard of the other, arising from
the internal frame and constitution of animals, is ultimately
derived from the Supreme Will, which bestowed on each being
its peculiar nature, and arranged the several classes and orders
of existence."—(IV. pp. 376—7.)

Hume has not discussed the theological theory

of the obligations of morality, but it is obviously in accordance with his view of the nature of those obligations. Under its theological aspect, morality is obedience to the will of God; and the ground for such obedience is two-fold : either we ought to obey God because He will punish us if we disobey Him, which is an argument based on the utility of obedience; or our obedience ought to flow from our love towards God, which is an argument based on pure feeling and for which no reason can be given. For, if any man should say that he takes no pleasure in the contemplation of the ideal of perfect holiness, or, in other words, that he does not love God, the attempt to argue him into acquiring that pleasure would be as hopeless as the endeavour to persuade Peter Bell of the "witchery of the soft blue sky."

In whichever way we look at the matter, morality is based on feeling, not on reason; though reason alone is competent to trace out the effects of our actions and thereby dictate conduct. Justice is founded on the love of one's neighbour; and goodness is a kind of beauty. The moral law, like the laws of physical nature, rests in the long run upon instinctive intuitions, and is neither more nor less "innate" and "necessary" than they are. Some people cannot by any means be got to understand the first book of Euclid ; but the truths of mathematics are no less necessary and binding on the great mass of mankind. Some there are who cannot feel the difference between

the "Sonata Appassionata" and "Cherry Ripe;"
or between a grave-stone-cutter's cherub and the
Apollo Belvidere ; but the canons of art are none
the less acknowledged. While some there may
be, who, devoid of sympathy, are incapable of a
sense of duty ; but neither does their existence
affect the foundations of morality. Such patho-
logical deviations from true manhood are merely
the halt, the lame, and the blind of the world of
consciousness ; and the anatomist of the mind
leaves them aside, as the anatomist of the body
would ignore abnormal specimens.

And as there are Pascals and Mozarts, Newtons
and Raffaelles, in whom the innate faculty for
science or art seems to need but a touch to spring
into full vigour, and through whom the human
race obtains new possibilities of knowledge and
new conceptions of beauty : so there have been
men of moral genius, to whom we owe ideals of
duty and visions of moral perfection, which
ordinary mankind could never have attained :
though, happily for them, they can feel the beauty
of a vision, which lay beyond the reach of their
dull imaginations, and count life well spent in
shaping some faint image of it in the actual world.

HELPS TO THE STUDY

OF

BERKELEY

BISHOP BERKELEY ON THE META-PHYSICS OF SENSATION [1]

[1871]

PROFESSOR FRASER has earned the thanks of all students of philosophy for the conscientious labour which he has bestowed upon his new edition of the works of Berkeley; in which, for the first time, we find collected together every thought which can be traced to the subtle and penetrating mind of the famous Bishop of Cloyne; while the "Life and Letters" will rejoice those who care less for the idealist and the prophet of tar-water, than for the man who stands out as one of the noblest and purest figures of his time: that Berkeley from whom the jealousy of Pope

---

[1] *The Works of George Berkeley, D.D., formerly Bishop of Cloyne, including many of his Works hitherto unpublished, with Preface, Annotations, his Life and Letters, and an Account of his Philosophy.* By A. C. Fraser. Four vols. Oxford: Clarendon Press, 1871.

R 2

did not withhold a single one of all "the virtues under heaven;" nor the cynicism of Swift, the dignity of "one of the first men of the kingdom for learning and virtue;" the man whom the pious Atterbury could compare to nothing less than an angel; whose personal influence and eloquence filled the Scriblerus Club and the House of Commons with enthusiasm for the evangelization of the North American Indians; and even led Sir Robert Walpole to assent to the appropriation of public money to a scheme which was neither business nor bribery.[1]

Hardly any epoch in the intellectual history of England is more remarkable in itself, or possesses a greater interest for us in these latter days, than that which coincides broadly with the conclusion of the seventeenth and the opening of the eighteenth century. The political fermentation of the preceding age was gradually working itself out; domestic peace gave men time to think; and the toleration won by the party of which Locke was the spokesman, permitted a freedom of speech and of writing such as has rarely been exceeded in later times. Fostered by these circumstances, the great faculty for physical and metaphysical

---

[1] In justice to Sir Robert, however, it is proper to remark that he declared afterwards, that he gave his assent to Berkeley's scheme for the Bermuda University only because he thought the House of Commons was sure to throw it out.

inquiry, with which the people of our race are naturally endowed, developed itself vigorously; and at least two of its products have had a profound and a permanent influence upon the subsequent course of thought in the world. The one of these was English Freethinking; the other, the Theory of Gravitation.

Looking back to the origin of the intellectual impulses of which these were the results, we are led to Herbert, to Hobbes, to Bacon; and to one who stands in advance of all these, as the most typical man of his time—Descartes. It is the Cartesian doubt—the maxim that assent may properly be given to no propositions but such as are perfectly clear and distinct—which, becoming incarnate, so to speak, in the Englishmen, Anthony Collins, Toland, Tindal, Woolston, and in the wonderful Frenchman, Pierre Bayle, reached its final term in Hume. And, on the other hand, although the theory of Gravitation set aside the Cartesian vortices—yet the spirit of the " Principes de Philosophie " attained its apotheosis when Newton demonstrated all the host of heaven to be but the elements of a vast mechanism, regulated by the same laws as those which express the falling of a stone to the ground. There is a passage in the preface to the first edition of the " Principia " which shows that Newton was penetrated, as completely as Descartes, with the belief that all the phenomena of nature[1] are expressible in terms of matter and motion.

"Would that the rest of the phenomena of nature could be deduced by a like kind of reasoning from mechanical principles. For many circumstances lead me to suspect that all these phenomena may depend upon certain forces, in virtue of which the particles of bodies, by causes not yet known, are either mutually impelled against one another and cohere into regular figures, or repel and recede from one another; which forces being unknown, philosophers have as yet explored nature in vain. But I hope that, either by this method of philosophizing, or by some other and better, the principles here laid down may throw some light upon the matter."[2]

[1] So far as Descartes is concerned the phenomena of consciousness are excluded from this category. According to his view, animals and man, in so far as he resembles them, are mechanisms. The soul, which alone feels and thinks, is extra-natural—a something divinely created and added to the anthropoid mechanism. He thus provided their favourite resting-place for the supranaturalistic evolutionists of our day.

Descartes' denial of sensation to the lower animals is a necessary consequence of his hypothesis concerning the nature and origin of the soul. He was too logical a thinker not to be aware that, if he admitted even the most elementary form of consciousness to be a product or a necessary concomitant, of material mechanism, the assumption of the existence of a thinking substance, apart from matter, would become superfluous.—[1894].

[2] "Utinam cætera naturæ phænomena ex principiis mechanicis, eodem argumentandi genere, derivare licet. Nam multa me movent, ut nonnihil suspicer ea omnia ex viribus quibusdam pendere posse, quibus corporum particulæ, per causas nondum cognitas, vel in se mutuo impelluntur et secundum figuras

But the doctrine that all the phenomena of
nature are resolvable into mechanism is what
people have agreed to call "materialism;" and
when Locke and Collins maintained that matter
may possibly be able to think, and Newton
himself could compare infinite space to the sensorium of the Deity, it was not wonderful that
the English philosophers should be attacked as
they were by Leibnitz in the famous letter to the
Princess of Wales, which gave rise to his correspondence with Clarke.[1]

"1. Natural religion itself seems to decay [in
England] very much. Many will have human
souls to be material; others make God Himself a
corporeal Being.

"2. Mr. Locke and his followers are uncertain,
at least, whether the soul be not material and
naturally perishable.

"3. Sir Isaac Newton says that space is an
organ which God makes use of to perceive things
by. But if God stands in need of any organ to
perceive things by, it will follow that they do not
depend altogether upon Him, nor were produced
by Him.

regulares cohærent vel ab invicem fugantur et recedunt; quibus
viribus ignotis, Philosophi hactenus Naturam frustra tentarunt.
Spero autem qucd vel huic philosophandi modo, vel veriori,
alicui, principia hic posita lucem aliquam præbebunt."—Preface
to First Edition of *Principia*, May 8, 1686.
    [1] *Collection of Papers which passed between the learned late*
Mr. Leibnitz and Dr. Clarke.—1717.

"4. Sir Isaac Newton and his followers have also a very odd opinion concerning the work of God. According to their doctrine, God Almighty wants to wind up His watch from time to time; otherwise it would cease to move.[1] He had not, it seems, sufficient foresight to make it a perpetual motion. Nay, the machine of God's making is so imperfect, according to these gentlemen, that He is obliged to clean it now and then by an extraordinary concourse, and even to mend it as a clockmaker mends his work."

It is beside the mark, at present, to inquire how far Leibnitz paints a true picture, and how far he is guilty of a spiteful caricature of Newton's views in these passages; and whether the beliefs which Locke is known to have entertained are consistent with the conclusions which may logically be drawn from some parts of his works. It is undeniable that English philosophy in Leibnitz's time had the general character which he ascribes to it. The phenomena of nature were held to be resolvable into the attractions and the repulsions of particles of matter; all knowledge was attained through the senses; the mind antecedent to experience was a *tabula rasa*. In other words, at the commencement of the eighteenth century, the character of speculative thought in

---

[1] Goethe seems to have had this saying of Leibnitz in his mind when he wrote his famous lines—
   " Was wär' ein Gott der nur von aussen stiesse
   Im Kreis das All am Finger laufen liesse."

England was essentially sceptical, critical, and materialistic. Why such "materialism" [1] should be more inconsistent with the existence of a Deity, the freedom of the will, or the immortality of the soul, or with any actual or possible system of theology, than "idealism," I must declare myself at a loss to divine. But, in the year 1700, all the world appears to have been agreed, Tertullian notwithstanding, that materialism necessarily leads to very dreadful consequences. And it was thought that it conduced to the interests of religion and morality to attack the materialists with all the weapons that came to hand. Perhaps the most interesting controversy which arose out of these questions is the wonderful triangular duel between Dodwell, Clarke, and Anthony Collins, concerning the materiality of the soul, and—what all the disputants considered to be the necessary consequence of its materiality—its natural mortality. I do not think that any one can read the letters which passed between Clarke and Collins, without admitting that Collins, who writes with wonderful power and closeness of reasoning, has by far the best of the argument, so far as the possible materiality of the soul goes; and that, in this battle, the Goliath of Freethinking overcame the champion of what was considered Orthodoxy.

In Dublin, all this while, there was a little

---

[1] *See* Note A appended to this Essay.

David practising his youthful strength upon the intellectual lions and bears of Trinity College. This was George Berkeley, who was destined to give the same kind of development to the idealistic side of Descartes' philosophy, that the Freethinkers had given to its sceptical side, and the Newtonians to its mechanical side.

Berkeley faced the problem boldly. He said to the materialists: "You tell me that all the phenomena of nature are resolvable into matter and its affections. I assent to your statement, and now I put to you the further question, 'What is matter?' In answering this question you shall be bound by your own conditions; and I demand, in the terms of the Cartesian axiom, that in turn you give your assent only to such conclusions as are perfectly clear and obvious."

It is this great argument which is worked out in the "Treatise concerning the Principles of Human Knowledge," and in those "Dialogues between Hylas and Philonous," which rank among the most exquisite examples of English style, as well as among the subtlest of metaphysical writings; and the final conclusion of which is summed up in a passage remarkable alike for literary beauty, and for calm audacity of statement.

"Some truths there are so near and obvious to the mind that a man need only open his eyes to see them. Such I take this important one to be, viz., that all the choir of heaven and furniture of the earth—in a word, all those bodies which compose the mighty frame of the world—have not any substance without a mind ; that their being is to be perceived or known ; that consequently, so long as they are not actually perceived by me, or do not exist in my mind or that of any other created spirit, they must either have no existence at all or else subsist in the mind of some eternal spirit ; it being perfectly unintelligible, and involving all the absurdity of abstraction, to attribute to any single part of them an existence independent of a spirit." [1]

Doubtless this passage sounds like the acme of metaphysical paradox, and we all know that "coxcombs vanquished Berkeley with a grin;" while common-sense folk refuted him by stamping on the ground, or some such other irrelevant proceeding. But the key to all philosophy lies in the clear apprehension of Berkeley's problem— which is neither more nor less than one of the shapes of the greatest of all questions, "What are the limits of our faculties?" And it is worth any amount of trouble to comprehend the exact nature of the argument by which Berkeley arrived at his results, and to know by one's own knowledge the great truth which he discovered—that the honest and rigorous following up of the argument which leads us to "materialism," inevitably carries us beyond it.

Suppose that I accidentally prick my finger with a pin. I immediately become aware of a

[1] *Treatise concerning the Principles of Human Knowledge*, Part I. § 6.

condition of my consciousness—a feeling which I term pain. I have no doubt whatever that the feeling is in myself alone; and if any one were to say that the pain I feel is something which inheres in the needle, as one of the qualities of the substance of the needle, we should all laugh at the absurdity of the phraseology. In fact, it is utterly impossible to conceive pain except as a state of consciousness.

Hence, so far as pain is concerned, it is sufficiently obvious that Berkeley's phraseology is strictly applicable to our power of conceiving its existence—"its being is to be perceived or known," and "so long as it is not actually perceived by me, or does not exist in my mind, or that of any other created spirit, it must either have no existence at all, or else subsist in the mind of some eternal spirit."

So much for pain. Now let us consider an ordinary sensation. Let the point of the pin be gently rested upon the skin, and I become aware of a feeling, or condition of consciousness, quite different from the former—the sensation of what I call "touch." Nevertheless this touch is plainly just as much in myself as the pain was. I cannot for a moment conceive this something which I call touch as existing apart from myself, or a being capable of the same feelings as myself. And the same reasoning applies to all the other simple sensations. A moment's reflection is sufficient to convince one that the smell, and the taste, and the yellowness, of which we become aware when an orange is smelt, tasted, and seen, are as completely states of our consciousness as is the pain which arises if the orange happens to be too sour. Nor is it less clear that every sound is a state of the consciousness of him who hears it. If the universe contained only blind and deaf beings, it is impossible for us to imagine but that darkness and silence should reign everywhere.

It is undoubtedly true, then, of all the simple sensations that, as Berkeley says, their "*esse* is *percipi*"—their being is to be "perceived or known." But that which perceives, or knows, is termed mind or spirit; and therefore the knowledge which the senses give us is, after all, a knowledge of spiritual phenomena.

All this was explicitly or implicitly admitted, and, indeed, insisted upon, by Berkeley's contemporaries, and by no one more strongly than by Locke, who terms smells, tastes, colours, sounds, and the like, "secondary qualities," and observes, with respect to these "secondary qualities," that "whatever reality we by mistake attribute to them [they] are in truth nothing in the objects themselves."

And again: "Flame is denominated hot and light; snow, white and cold; and manna, white and sweet, from the ideas they produce in us; which qualities are commonly thought to be the same in these bodies; that those ideas are in us, the one the perfect resemblance of the other as they are in a mirror; and it would by most men be judged very extravagant if one should say otherwise. And yet he that will consider that the same fire that at one distance produces in us the sensation of warmth, does at a nearer approach produce in us the far different sensation of pain, ought to bethink himself what reason he has to say that his idea of warmth, which was produced in him by the fire, is actually in the fire; and his idea of pain which the same fire produced in him in the same way, is not in the fire. Why are whiteness and coldness in snow, and pain not, when it produces the one and the other idea in us; and can do neither but by the bulk, figure, number, and motion of its solid parts?"[1]

Thus far then materialists and idealists are agreed. Locke and Berkeley, and all logical thinkers who have succeeded them, are of one mind about secondary qualities—their being is to be perceived or known—their materiality is, in strictness, a spirituality.

But Locke draws a great distinction between the secondary qualities of matter, and certain others which he terms "primary qualities." These are extension, figure, solidity, motion and rest, and number; and he is as clear that these

[1] Locke, *Human Understanding*, Book II. chap. viii. §§ 14, 15.

primary qualities exist independently of the mind, as he is that the secondary qualities have no such existence.

"The particular bulk, number, figure, and motion of the parts of fire and snow are really in them, whether any one's senses perceive them or not, and therefore they may be called real qualities, because they really exist in those bodies ; but light, heat, whiteness, or coldness, are no more really in them, than sickness, or pain, is in manna. Take away the sensation of them ; let not the eyes see light or colours, nor the ears hear sounds ; let the palate not taste, nor the nose smell ; and all colours, tastes, odours and sounds, as they are such particular ideas, vanish and cease, and are reduced to their causes, *i.e.* bulk, figure, and motion of parts.

"18. A piece of manna of sensible bulk is able to produce in us the idea of a round or square figure ; and, by being removed from one place to another, the idea of motion. This idea of motion represents it as it really is in the manna moving ; a circle and square are the same, whether in idea or existence, in the mind or in the manna ; and thus both motion and figure are really in the manna, whether we take notice of them or no : this everybody is ready to agree to."

So far as primary qualities are concerned, then, Locke is as thoroughgoing a realist as St. Anselm. In Berkeley, on the other hand, we have as complete a representative of the nominalists and conceptualists—an intellectual descendant of Roscellinus and of Abelard.[1] And by a curious irony of fate, it is the nominalist who is, this time, the champion of orthodoxy, and the realist that of heresy.

Once more let us try to work out Berkeley's

[1] See note B

principles for ourselves, and inquire what foundation there is for the assertion that extension, form, solidity, and the other "primary qualities," have an existence apart from mind. And for this purpose let us recur to our experiment with the pin.

It has been seen that when the finger is pricked with a pin, a state of consciousness arises which we call pain; and it is admitted that this pain is not a something which inheres in the pin, but a something which exists only in the mind, and has no similitude elsewhere.

But a little attention will show that this state of consciousness is accompanied by another, which can by no effort be got rid of. I not only have the feeling, but the feeling is localized. I am just as certain that the pain is in my finger, as I am that I have it at all. Nor will any effort of the imagination enable me to believe that the pain is not in my finger.

And yet nothing is more certain than that it is not, and cannot be, in the spot in which I feel it, nor within a couple of feet of that spot. For the skin of the finger is connected by a bundle of fine nervous fibres, which run up the whole length of the arm, to the spinal marrow, which sets them in communication with the brain, and we know that the feeling of pain caused by the prick of a pin is dependent on the integrity of those fibres. After they have been cut through close to the spinal cord, no pain will be felt, whatever injury

is done to the finger; and if the ends which remain in connection with the cord be pricked, the
pain which arises will appear to have its seat in
the finger just as distinctly as before. Nay, if the
whole arm be cut off, the pain which arises from
pricking the nerve stump will appear to be seated
in the fingers, just as if they were still connected
with the body.

It is perfectly obvious, therefore, that the
localization of the pain at the surface of the body
is an act of the mind. It is an *extradition* of
that consciousness, which has its seat in the
brain, to a definite point of the body—which
takes place without our volition, and may give
rise to ideas which are contrary to fact. We
might call this extradition of consciousness a
reflex feeling, just as we speak of a movement
which is excited apart from, or contrary to, our
volition, as a reflex motion. Locality is no more
in the pin than pain is; of the former, as of the
latter, it is true that "its being is to be perceived," and that its existence apart from a
thinking mind is not conceivable.

The foregoing reasoning will be in no way
affected, if, instead of pricking the finger, the
point of the pin rests gently against it, so as to
give rise merely to a tactile sensation. The
tactile sensation is referred outwards to the point
touched, and seems to exist there. But it is
certain that it is not and cannot be there really,

because the brain is the sole seat of consciousness; and, further, because evidence, as strong as that in favour of the sensation being in the finger, can be brought forward in support of propositions which are manifestly absurd. For example, the hairs and nails are utterly devoid of sensibility, as every one knows. Nevertheless, if the ends of the nails or hairs are touched, ever so lightly, we feel that they are touched, and the sensation seems to be situated in the nails or hairs. Nay more, if a walking-stick, a yard long, is held firmly by the handle and the other end is touched, the tactile sensation, which is a state of our own consciousness, is unhesitatingly referred to the end of the stick; and yet no one will say that it *is* there.

Let us now suppose that, instead of one pin's point resting against the end of my finger, there are two. Each of these can be known to me, as we have seen, only as a state of a thinking mind, referred outwards, or localized. But the existence of these two states, somehow or other, generates in my mind a number of new ideas, which did not make their appearance when only one state was present. For example, I get the ideas of co-existence, of number, of distance, and of relative place or direction. But all these ideas are ideas of relations, and may be said to imply the existence of something which perceives those relations. If a tactile sensation is a state of the mind, and if

the localization of that sensation is an act of the
mind, how is it conceivable that a relation between two localized sensations should exist apart
from the mind ? It is, I confess, quite as easy
for me to imagine that redness may exist apart
from a visual sense, as it is to suppose that
co-existence, number, and distance can have any
existence apart from the mind of which they are
ideas.

Thus it seems clear that the existence of some,
at any rate, of Locke's primary qualities of matter,
such as number and extension, apart from mind,
is as utterly unthinkable as the existence of colour
and sound under like circumstances.

Will the others—namely, figure, motion and
rest, and solidity—withstand a similar criticism ?
I think not. For all these, like the foregoing, are
perceptions by the mind of the relations of two or
more sensations to one another. If distance and
place are inconceivable, in the absence of the mind
of which they are ideas, the independent existence
of figure, which is the limitation of distance, and of
motion, which is change of place, must be equally
inconceivable. Solidity requires more particular
consideration, as it is a term applied to two very
different things, the one of which is solidity of
form, or geometrical solidity; while the other is
solidity of substance, or mechanical solidity. If
those motor nerves of a man by which volitions
are converted into motion were all paralysed, and

s 2

if sensation remained only in the palm of his hand (which is a conceivable case), he would still be able to attain to clear notions of extension, figure, number, and motion by attending to the states of consciousness which might be aroused by the contact of bodies with the sensory surface of the palm. But it does not appear that such a person could arrive at any conception of geometrical solidity. For that which does not come in contact with the sensory surface is non-existent for the sense of touch ; and a solid body, impressed upon the palm of the hand, gives rise only to the notion of the extension of that particular part of the solid which is in contact with the skin.

Nor is it possible that the idea of outness (in the sense of discontinuity with the sentient body) could be attained by such a person; for, as we have seen, every tactile sensation is referred to a point either of the natural sensory surface itself, or of some solid in continuity with that surface. Hence it would appear that the conception of the difference between the Ego and the non-Ego could not be attained by a man thus situated. His feelings would be his universe, and his tactile sensations his "mœnia mundi." Time would exist for him as for us, but space would have only two dimensions.

But now remove the paralysis from the motor apparatus, and give the palm of the hand of our imaginary man perfect freedom to move, so as to be able to glide in all directions over the bodies with which it is in contact. Then with the consciousness of that mobility, the notion of space of three dimensions—which is " *Raum*," or " room " to move with perfect freedom—is at once given. But the notion that the tactile surface itself moves, cannot be given by touch alone, which is competent to testify only to the fact of change of place, not to its cause. The idea of the motion of the tactile surface could not, in fact, be attained, unless the idea of change of place were accompanied by some state of consciousness, which does not exist when the tactile surface is immoveable. This state of consciousness is what is termed the muscular sense, and its existence is very easily demonstrable.

Suppose the back of my hand to rest upon a table, and a sovereign to rest upon the upturned palm, I at once acquire a notion of extension, and of the limit of that extension. The impression made by the circular piece of gold is quite different from that which would be made by a triangular, or a square, piece of the same size, and thereby I arrive at the notion of figure. Moreover, if the sovereign slides over the palm, I acquire a distinct conception of change of place or motion, and of the direction of that motion. For as the sovereign slides, it affects new nerve-endings, and gives rise to new states of consciousness. Each of them is definitely and separately localized by a reflex act of the mind, which, at the same time, becomes aware of the difference between two successive localizations; and therefore of change of place, which is motion.

If, while the sovereign lies on the hand, the latter being kept quite steady, the fore-arm is gradually and slowly raised; the tactile sensations, with all their accompaniments, remain exactly as they were. But, at the same time, something new is introduced; namely, the sense of effort. If I try to discover where this sense of effort seems to be, I find myself somewhat perplexed at first; but, if I hold the fore-arm in position long enough, I become aware of an obscure sense of fatigue, which is apparently seated either in the muscles of the arm, or in the integument directly over them. The fatigue seems to be related to the sense of effort, in much the same way as the pain which supervenes upon the original sense of contact, when a pin is slowly pressed against the skin, is related to touch.

A little attention will show that this sense of effort accompanies every muscular contraction by which the limbs, or other parts of the body, are moved. By its agency the fact of their movement is known; while the direction of the motion is given by the accompanying tactile sensations. And, in consequence of the incessant association of the muscular and the tactile sensations, they become so fused together that they are often confounded under the same name.

If freedom to move in all directions is the very essence of that conception of space of three dimensions which we obtain by the sense of touch; and if that freedom to move is really another name for the feeling of unopposed effort, accompanied by that of change of place, it is surely impossible to conceive of such space as having existence apart from that which is conscious of effort.

But it may be said that we derive our conception of space of three dimensions not only from touch, but from vision; that if we do not feel things actually outside us, at any rate we see them. And it was exactly this difficulty which presented itself to Berkeley at the outset of his speculations. He met it, with characteristic boldness, by denying that we do see things outside us; and, with no less characteristic ingenuity, by devising that " New Theory of Vision " which has met with wider acceptance than any of his views, though it has been the subject of continual controversies.[1]

In the " Principles of Human Knowledge," Berkeley himself tells us how he was led to those

[1] I have not specifically alluded to the writings of Bailey, Mill, Abbott, and others, on this vexed question, not because I have failed to study them carefully, but because this is not a convenient occasion for controversial discussion. Those who are acquainted with the subject, however, will observe that the view I have taken agrees substantially with that of Mr. Bailey.

opinions which he published in the "Essay towards the New Theory of Vision."

"It will be objected that we see things actually without, or at a distance from us, and which consequently do not exist in the mind ; it being absurd that those things which are seen at the distance of several miles, should be as near to us as our own thoughts. In answer to this, I desire it may be considered that in a dream we do oft perceive things as existing at a great distance off, and yet, for all that, those things are acknowledged to have their existence only in the mind.

"But for the fuller clearing of this point, it may be worth while to consider how it is that we perceive distance and things placed at a distance by sight. For that we should in truth see external space and bodies actually existing in it, some nearer, others further off, seems to carry with it some opposition to what hath been said of their existing nowhere without the mind. The consideration of this difficulty it was that gave birth to my "Essay towards the New Theory of Vision" which was published not long since, wherein it is shown that distance, or outness, is neither immediately of itself perceived by sight, nor yet apprehended, or judged of, by lines and angles or anything that hath any necessary connection with it ; but that it is only suggested to our thoughts by certain visible ideas and sensations attending vision, which, in their own nature, have no manner of similitude or relation either with distance or with things placed at a distance ; but by a connection taught us by experience, they come to signify and suggest them to us, after the same manner that words of any language suggest the ideas they are made to stand for ; insomuch that a man born blind and afterwards made to see, would not, at first sight, think the things he saw to be without his mind or at any distance from him."

The key-note of the Essay to which Berkeley refers in this passage is to be found in an italicized paragraph of section 127 :—

" The extensions, figures, and motions perceived by sight are specifically distinct from the ideas of touch called by the same names; nor is there any such thing as an idea or kind of idea common to both senses."

It will be observed that this proposition expressly declares that extension, figure, and motion, and consequently distance, are immediately perceived by sight as well as by touch; but that visual distance, extension, figure, and motion, are totally different in quality from the ideas of the same name obtained through the sense of touch. And other passages leave no doubt that such was Berkeley's meaning. Thus in the 112th section of the same Essay, he carefully defines the two kinds of distance, one visual, the other tangible :—

"By the distance between any two points nothing more is meant than the number of intermediate points. If the given points are visible, the distance between them is marked out by the number of interjacent visible points; if they are tangible, the distance between them is a line consisting of tangible points."

Again, there are two sorts of magnitude or extension :—

"It has been shown that there are two sorts of objects apprehended by sight, each whereof has its distinct magnitude or extension : the one properly tangible, *i.e.*, to be perceived and measured by touch, and not immediately falling under the sense of seeing; the other properly and immediately visible, by mediation of which the former is brought into view."—§ 55.

But how are we to reconcile these passages with others which will be perfectly familiar to every

reader of the " New Theory of Vision " ?　As, for
example :—

"It is, I think, agreed by all, that distance of itself, and
immediately, cannot be seen."—§ 2.

"Space or distance, we have shown, is no otherwise the object
of sight than of hearing."—§ 130.

"Distance is in its own nature imperceptible, and yet it is
perceived by sight.　It remains, therefore, that it is brought
into view by means of some other idea, that is itself immediately
perceived in the act of vision."—§ 11.

"Distance or external space."—§ 155.

The explanation is quite simple, and lies in the
fact that Berkeley uses the word "distance" in
three senses.　Sometimes he employs it to denote
visible distance, and then he restricts it to distance
in two dimensions, or simple extension.　Sometimes he means tangible distance in two dimensions; but most commonly he intends to signify
tangible distance in the third dimension.　And it
is in this sense that he employs "distance" as the
equivalent of "space."　Distance in two dimensions is, for Berkeley, not space, but extension.
By taking a pencil and interpolating the words
"visible" and "tangible" before "distance"
wherever the context renders them necessary,
Berkeley's statements may be made perfectly consistent; though he has not always extricated himself from the entanglement caused by his own
loose phraseology, which rises to a climax in the
last ten sections of the "Theory of Vision," in
which he endeavours to prove that a pure intelli-

gence able to see, but devoid of the sense of touch, could have no idea of a plane figure. Thus he says in section 156 :—

"All that is properly perceived by the visual faculty amounts to no more than colours with their variations and different proportions of light and shade ; but the perpetual mutability and fleetingness of those immediate objects of sight render them incapable of being managed after the manner of geometrical figures, nor is it in any degree useful that they should. It is true there be divers of them perceived at once, and more of some and less of others ; but accurately to compute their magnitude, and assign precise determinate proportions between things so variable and inconstant, if we suppose it possible to be done, must yet be a very trifling and insignificant labour."

If, by this, Berkeley means that by vision alone, a straight line cannot be distinguished from a curved one, a circle from a square, a long line from a short one, a large angle from a small one, his position is surely absurd in itself and contradictory to his own previously cited admissions; if he only means, on the other hand, that his pure spirit could not get very far on in his geometry, it may be true or not; but it is in contradiction with his previous assertion, that such a pure spirit could never attain to know as much as the first elements of plane geometry.

Another source of confusion, which arises out of Berkeley's insufficient exactness in the use of language, is to be found in what he says about solidity, in discussing Molyneux's problem, whether a man born blind and having learned to distinguish between a cube and a sphere, could, on receiving his sight, tell the one from the other by vision. Berkeley agrees with Locke that he could not, and adds the following reflection:—

"Cube, sphere, table, are words he has known applied to things perceivable by touch, but to things perfectly intangible he never knew them applied. Those words in their wonted application always marked out to his mind bodies or solid things which were perceived by the resistance they gave. But there is no solidity, no resistance or protrusion perceived by sight."

Here "solidity" means resistance to pressure, which is apprehended by the muscular sense; but when in section 154 Berkeley says of his pure intelligence—

"It is certain that the aforesaid intelligence could have no idea of a solid or quantity of three dimensions, which follows from its not having any idea of distance"—

he refers to that notion of solidity which may be obtained by the tactile sense without the addition of any notion of resistance in the solid object; as, for example, when the finger passes lightly over the surface of a billiard ball.

Yet another source of difficulty in clearly understanding Berkeley arises out of his use of the word "outness." In speaking of touch he seems to employ it indifferently, both for the localization of a tactile sensation in the sensory surface, which we really obtain through touch; and for the notion of corporeal separation, which is attained by the association of muscular and tactile sensations. In speaking of sight, on the other hand, Berkeley employs "outness" to denote corporeal separation.

When due allowance is made for the occasional looseness and ambiguity of Berkeley's terminology, and the accessories are weeded out of the essential parts of his famous Essay, his views may, I believe, be fairly and accurately summed up in the following propositions :—

1. The sense of touch gives rise to ideas of extension, figure, magnitude, and motion.

2. The sense of touch gives rise to the idea of "outness," in the sense of localization.

3. The sense of touch gives rise to the idea of resistance, and thence to that of solidity, in the sense of impenetrability.

4. The sense of touch gives rise to the idea of "outness," in the sense of distance in the third dimension, and thence to that of space or geometrical solidity.

5. The sense of sight gives rise to ideas of extension, of figure, magnitude, and motion.

6. The sense of sight does not give rise to the idea of "outness," in the sense of distance in the third dimension, nor to that of geometrical solidity, no visual idea appearing to be without the mind, or at any distance off (§ § 43, 50).

7. The sense of sight does not give rise to the idea of mechanical solidity.

8. There is no likeness whatever between the

tactile ideas called extension, figure, magnitude, and motion, and the visual ideas which go by the same names; nor are any ideas common to the two senses.

9. When we think we see objects at a distance, what really happens is that the visual picture suggests that the object seen has tangible distance; we confound the strong belief in the tangible distance of the object with actual sight of its distance.

10. Visual ideas, therefore, constitute a kind of language, by which we are informed of the tactile ideas which will, or may, arise in us.

Taking these propositions into consideration *seriatim*, it may be assumed that every one will assent to the first and second; and that for the third and fourth we have only to include the muscular sense under the name of sense of touch, as Berkeley did, in order to make it quite accurate. Nor is it intelligible to me that any one should explicitly deny the truth of the fifth proposition, though some of Berkeley's supporters, less careful than himself, have done so. Indeed, it must be confessed that it is only grudgingly, and as it were against his will, that Berkeley admits that we obtain ideas of extension, figure, and magnitude by pure vision, and that he more than half retracts the admission; while he absolutely denies that sight gives us any notion of outness in either sense of the word, and even declares that "no proper visual idea appears to be without the mind,

or at any distance off." By "proper visual ideas,"
Berkeley denotes colours, and light, and shade;
and, therefore, he affirms that colours do not
appear to be at any distance from us. I confess
that this assertion appears to me to be utterly
unaccountable. I have made endless experiments
on this point, and by no effort of the imagination
can I persuade myself, when looking at a colour,
that the colour is in my mind, and not at a
"distance off," though of course I know perfectly
well, as a matter of reason, that colour is subjective. It is like looking at the sun setting, and
trying to persuade one's self that the earth appears
to move and not the sun, a feat I have never been
able to accomplish. Even when the eyes are
shut, the darkness of which one is conscious, carries
with it the notion of outness. One looks, so to
speak, into a dark space. Common language expresses the common experience of mankind in this
matter. A man will say that a smell is in his nose,
a taste is in his mouth, a singing is in his ears, a
creeping or a warmth is in his skin; but if he is
jaundiced, he does not say that he has yellow in
his eyes, but that everything looks yellow; and if
he is troubled with *muscæ volitantes*, he says, not
that he has specks in his eyes, but that he sees
specks dancing before his eyes. In fact, it appears
to me that it is the special peculiarity of visual
sensations, that they invariably give rise to the
idea of remoteness, and that Berkeley's dictum

ought to be reversed. For I think that any one who interrogates his consciousness carefully will find that "every proper visual idea" appears to be without the mind and at a distance off.

Not only does every *visibile* appear to be remote, but it has a position in external space, just as a *tangibile* appears to be superficial and to have a determinate position on the surface of the body. Every *visibile*, in fact, appears (approximately) to be situated upon a line drawn from it to the point of the retina on which its image falls. It is referred outwards, in the general direction of the pencil of light by which it is rendered visible, just as, in the experiment with the stick, the *tangibile* is referred outwards to the end of the stick.

It is for this reason that an object, viewed with both eyes, is seen single and not double. Two distinct images are formed, but each image is referred to that point at which the two optic axes intersect; consequently, the two images cover one another, and appear as completely one as any other two equally similar super-imposed images would be.[1] And it is for the same reason, that, if the side of the ball of the eye is pressed upon at any point, a spot of light appears apparently outside the eye, and in a region exactly opposite to that in which the pressure is made.

But while it seems to me that there is no reason

[1] In the case of a near, solid, external object, such as a cube, this is not the whole story.

to doubt that the extradition of sensation is more complete in the case of the eye than in that of the skin, and that corporeal distinctness, and hence space, are directly suggested by vision, it is another, and a much more difficult question, whether the notion of geometrical solidity is attainable by pure vision; that is to say, by a single eye, all the parts of which are immoveable. However this may be for an absolutely fixed eye, I conceive there can be no doubt in the case of an eye that is moveable and capable of adjustment. For, with the moveable eye, the muscular sense comes into play in exactly the same way as with the moveable hand; and the notion of change of place, *plus* the sense of effort, gives rise to a conception of visual space, which runs exactly parallel with that of tangible space. When two moveable eyes are present, the notion of space of three dimensions is obtained in the same way as it is by the two hands, but with much greater precision.[1] And if, to take a case similar to one already assumed, we suppose a man deprived of every sense except vision, and of all motion except that of his eyes, it surely cannot be doubted that he would have a perfect conception of space; and indeed a much more perfect conception than he who possessed touch alone without vision. But of course our touchless man would be devoid of any notion of resistance; and hence space, for

[1] *See* Note C

him, would be altogether geometrical and devoid
of body.

And here another curious consideration arises,
what likeness, if any, would there be between the
visual space of the one man, and the tangible
space of the other ?

Berkeley, as we have seen (in the eighth proposition), declares that there is no likeness
between the ideas given by sight and those given
by touch ; and one cannot but agree with him, so
long as the term ideas is restricted to mere sensations. Obviously, there is no more likeness between the feel of a surface and the colour of it,
than there is between its colour and its smell.
All simple sensations, derived from different
senses, are incommensurable with one another,
and only gradations of their own intensity are
comparable. And thus, so far as the primary
facts of sensation go, visual figure and tactile
figure, visual magnitude and tactile magnitude,
visual motion and tactile motion, are truly unlike,
and have no common term. But when Berkeley
goes further than this, and declares that there are
no "ideas" common to the "ideas" of touch and
those of sight, it appears to me that he has fallen
into a great error, and one which is the chief
source of his paradoxes about geometry.

Berkeley in fact employs the word "idea," in
this instance, to denote two totally different classes
of feelings, or states of consciousness. For these

may be divided into two groups: the primary feelings, which exist in themselves and without relation to any other, such as pleasure and pain, desire, and the simple sensations obtained through the sensory organs; and the secondary feelings, which express those relations of primary feelings which are perceived by the mind; and the existence of which, therefore, implies the pre-existence of at least two of the primary feelings. Such are likeness and unlikeness in quality, quantity, or form succession and contemporaneity; contiguity and distance: cause and effect; motion and rest.

Now it is quite true that there is no likeness between the primary feelings which are grouped under sight and touch; but it appears to me wholly untrue, and indeed absurd, to affirm that there is no likeness between the secondary feelings which express the relations of the primary ones.

The relation of succession perceived between the visible taps of a hammer, is, to my mind, exactly like the relation of succession between the tangible taps; the unlikeness between red and blue is a mental phenomenon of the same order as the unlikeness between rough and smooth. Two points visibly distant are so, because one or more units of visible length (*minima visibilia*) are interposed between them; and as two points tangibly distant are so, because one or more units of tangible length (*minima tangibilia*) are interposed between them, it is clear that the notion of

T 2

interposition of units of sensibility, or *minima sensibilia*, is an idea common to the two. And whether I see a point move across the field of vision towards another point, or feel the like motion, the idea of the gradual diminution of the number of sensible units between the two points appears to me to be common to both kinds of motion.

Hence, I conceive, that though it be true that there is no likeness between the primary feelings given by sight and those given by touch, yet there is a complete likeness between the secondary feelings aroused by each sense.

Indeed, if it were not so, how could Logic, which deals with those forms of thought which are applicable to every kind of subject-matter, be possible? How could numerical proportion be as true of *visibilia*, as of *tangibilia*, unless there were some ideas common to the two? And to come directly to the heart of the matter, is there any more difference between the relations between tangible sensations which we call place and direction, and those between visible sensations which go by the same name, than there is between those relations of tangible and visible sensations which we call succession? And if there be none, why is Geometry not just as much a matter of *visibilia* as of *tangibilia*?

Moreover, as a matter of fact, it is certain that the muscular sense is so closely connected with both the visual and the tactile senses, that, by the ordinary laws of association, the ideas which it suggests must needs be common to both.

From what has been said it will follow that the ninth proposition falls to the ground; and that vision, combined with the muscular sensations produced by the movement of the eyes, gives us as complete a notion of corporeal separation and of distance in the third dimension of space, as touch, combined with the muscular sensations produced by the movements of the hand, does. The tenth proposition seems to contain a perfectly true statement, but it is only half the truth. It is no doubt true that our visual ideas are a kind of language by which we are informed of the tactile ideas which may or will arise in us; but this is true, more or less, of every sense in regard to every other. If I put my hand in my pocket, the tactile ideas which I receive prophesy quite accurately what I shall see—whether a bunch of keys or half-a-crown—when I pull it out again; and the tactile ideas are, in this case, the language which informs me of the visual ideas which will arise. So with the other senses: olfactory ideas tell me I shall find the tactile and visual phenomena called violets, if I look for them; taste, combined with touch, tells me that what I am tasting and touching with the tongue will, if I look at it, have the form of a clove; and hearing warns me of what I shall, or may, see and touch every minute of my life.

But while the " New Theory of Vision " cannot be considered to possess much value in relation to the immediate object its author had in view, it had a vastly important influence in directing attention to the real complexity of many of those phenomena of sensation, which appear at first to be simple.   And even if Berkeley, as I imagine, was quite wrong in supposing that we do not see space, the contrary doctrine makes quite as strongly for his general view, that space can be conceived only as something thought by a mind.

The last of Locke's " primary qualities " which remain to be considered is mechanical solidity, or impenetrability.   But our conception of this is derived from the sense of resistance to our own effort, or active force, which we meet with in association with sundry tactile or visual phenomena ; and, undoubtedly, active force is inconceivable except as a state of consciousness.   This may sound paradoxical; but let any one try to realize what he means by the mutual attraction of two particles, and I think he will find, either, that he conceives them simply as moving towards one another at a certain rate, in which case he only pictures motion to himself, and leaves force aside ; or, that he conceives each particle to be animated by something like his own volition, and to be pulling as he would pull.   And I suppose that this difficulty of thinking of force except as something comparable to volition lies at the bottom of

Leibnitz's doctrine of monads, to say nothing of Schopenhauer's "Welt als Wille und Vorstellung ;" while the opposite difficulty of conceiving force to be anything like volition, drives another school of thinkers into the denial of any connection, save that of succession, between cause and effect.

To sum up. If the materialist affirms that the universe and all its phenomena are resolvable into matter and motion, Berkeley replies, True; but what you call matter and motion are known to us only as forms of consciousness ; their being is to be conceived or known ; and the existence of a state of consciousness, apart from a thinking mind, is a contradiction in terms.

I conceive that this reasoning is irrefragable. And therefore, if I were obliged to choose between absolute materialism and absolute idealism, I should feel compelled to accept the latter alternative. Indeed, upon this point Locke does, practically, go as far in the direction of idealism as Berkeley, when he admits that "the simple ideas we receive from sensation and reflection are the boundaries of our thoughts, beyond which the mind, whatever efforts it would make, is not able to advance one jot."—Book II. chap. xxiii. § 29.

But Locke adds, "Nor can it make any discoveries when it would pry into the nature and hidden causes of these ideas."

Now, from this proposition, the thorough materialists dissent as much, on the one hand, as Berkeley does, upon the other hand.

The thorough materialist asserts that there is a something which he calls the "substance" of matter; that this something is the cause of all phenomena, whether material or mental; that it is self-existent and eternal, and so forth.

Berkeley, on the contrary, asserts, with equal confidence, that there is no substance of matter, but only a substance of mind, which he terms spirit; that there are two kinds of spiritual substance, the one eternal and uncreated, the substance of the Deity, the other created, and, once created, naturally eternal; that the universe, as known to created spirits, has no being in itself, but is the result of the action of the substance of the Deity on the substance of those spirits.

In contradiction to which bold assertion, Locke affirms that we simply know nothing about substance of any kind.[1]

"So that if any one will examine himself concerning his notion of pure substance in general, he will find he has no other idea of it at all, but only a supposition of he knows not what support of such qualities, which are capable of producing simple ideas in us, which qualities are commonly called accidents.

"If any one should be asked, what is the subject wherein

---

[1] Berkeley virtually makes the same confession of ignorance, when he admits that we can have no idea or notion of a spirit (*Principles of Human Knowledge*, § 138) ; and the way in which he tries to escape the consequences of this admission, is a splendid example of the floundering of a mired logician.

colour or weight inheres ? he would have nothing to say but the
solid extended parts ; and if he were demanded what is it that
solidity and extension inhere in ? he would not be in much
better case than the Indian before mentioned, who, urging that
the world was supported by a great elephant, was asked what
the elephant rested on ? to which his answer was, a great
tortoise. But being again pressed to know what gave support
to the broad-backed tortoise ? replied, something, he knew not
what. And thus here, as in all other cases when we use words
without having clear and distinct ideas, we talk like children,
who, being questioned what such a thing is, readily give this
satisfactory answer, that it is something ; which in truth signifies no more when so used, either by children or men, but that
they know not what, and that the thing they pretend to talk and
know of is what they have no distinct idea of at all, and are,
so, perfectly ignorant of it and in the dark. The idea, then,
we have, to which we give the general name substance, being
nothing but the supposed but unknown support of those
qualities we find existing, which we imagine cannot exist *sine
re substante*, without something to support them, we call that
support *substantia*, which, according to the true import of the
word, is, in plain English, standing under or upholding." [1]

I cannot but believe that the judgment of
Locke is that which Philosophy will accept as her
final decision.

Suppose that a rational piano were conscious of
sound, and of nothing else. It would be acquainted
with a system of nature entirely composed of
sounds, and the laws of nature would be the laws
of melody and of harmony. It might acquire
endless ideas of likeness and unlikeness, of
succession, of similarity and dissimilarity, but it

---

[1] Locke, *Human Understanding*, Book II. chap. xxiii. §2.

282 THE METAPHYSICS OF SENSATION

could attain to no conception of space, of distance, or of resistance; or of figure, or of motion.

The piano might then reason thus: All my knowledge consists of sounds and the perception of the relations of sounds; now the being of sound is to be heard; and it is inconceivable that the existence of the sounds I know, should depend upon any other existence than that of the mind of a hearing being.

This would be quite as good reasoning as Berkeley's, and very sound and useful, so far as it defines the limits of the piano's faculties. But for all that, pianos have an existence quite apart from sounds, and the auditory consciousness of our speculative piano would be dependent, in the first place, on the existence of a "substance" of brass, wood, and iron, and, in the second, on that of a musician. But of neither of these conditions of the existence of his consciousness would the phenomena of that consciousness afford him the slightest hint.

So that while it is the summit of human wisdom to learn the limit of our faculties, it may be wise to recollect that we have no more right to make denials, than to put forth affirmatives, about what lies beyond that limit. Whether either mind, or matter, has a "substance" or not, is a problem which we are incompetent to discuss; and it is just as likely that the common notions upon the subject should be correct as any others.

Indeed, Berkeley himself makes Philonous wind up his discussions with Hylas, in a couple of sentences which aptly express this conclusion :—

"You see, Hylas, the water of yonder fountain, how it is forced upwards in a round column to a certain height, at which it breaks and falls back into the basin from whence it rose ; its ascent as well as its descent proceeding from the same uniform law or principle of gravitation. Just so, the same principles which, at first view, lead to scepticism, pursued to a certain point, bring men back to common sense."

APPENDIX

NOTE A (p. 249.).

THE horror of "Materialism" which weighs upon the minds
of so many excellent people appears to depend, in part, upon the
purely accidental connexion of some forms of materialistic philosophy with ethical and religious tenets by which they are
repelled ; and, partly, on the survival of a very ancient superstition concerning the nature of matter.

This superstition, for the tenacious vitality of which the
idealistic philosophers who are, more or less, disciples of Plato
and the theologians who have been influenced by them, are
responsible, assumes that matter is something, not merely inert
and perishable, but essentially base and evil-natured, if not
actively antagonistic to, at least a negative dead-weight upon,
the good.   Judging by contemporary literature, there are
numbers of highly cultivated and indeed superior persons to
whom the material world is altogether contemptible ; who can
see nothing in a handful of garden soil, or a rusty nail, but
types of the passive and the corruptible.

To modern science, these assumptions are as much out of date
as the equally venerable errors, that the sun goes round the
earth every four-and-twenty hours, or that water is an elementary body.   The handful of soil is a factory thronged with
swarms of busy workers ; the rusty nail is an aggregation of
millions of particles, moving with inconceivable velocity in
a dance of infinite complexity yet perfect measure ; harmonic
with like performances throughout the solar system.   If there is
good ground for any conclusion, there is such for the belief that

the substance of these particles has existed and will exist, that the
energy which stirs them has persisted and will persist, without
assignable limit, either in the past or the future. Surely, as
Heracleitus said of his kitchen with its pots and pans, "Here
also are the gods." Little as we have, even yet, learned of the
material universe, that little makes for the belief that it is a
system of unbroken order and perfect symmetry, of which the
form incessantly changes, while the substance and the energy
are imperishable.

It will be understood that those who are thoroughly imbued
with this view of what is called "matter" find it a little
difficult to understand why that which is termed "mind"
should give itself such airs of superiority over the twin sister;
to whom, so far as our planet is concerned, it might be
hazardous to deny the right of primogeniture.

Accepting the ordinary view of mind, it is a substance the
properties of which are states of consciousness, on the one
hand, and energy of the same order as that of the material
world (or else it would not be able to affect the latter) on
the other hand. It is admitted that chance has no more place
in the world of mind, than it has in that of matter. Sensations,
emotions, intellections are subject to an order, as strict and inviolable as that which obtains among material things. If the order
which obtains in the material world lays it open to the reproach
of subjection to "blind necessity," the demonstrable existence
of a similar order amidst the phenomena of consciousness
(and without the belief in that fixed order, logic has no binding
force and morals have no foundation) renders it obnoxious to the
same condemnation. For necessity is necessity, and whether it
is blind or sharp-eyed is nothing to the purpose.

Even if the supposed energy of the substance of mind is
sometimes exerted without any antecedent cause—which is the
only intelligible sense of the popular doctrine of free-will—the
occurrence is admittedly exceptional, and, by the nature of
the case, it is not susceptible of proof. Moreover, if the hypothetical substance of mind is possessed of energy, I, for my
part, am unable to see how it is to be discriminated from the
hypothetical substance of matter.

Thus, if any man think he has reason to believe that the "substance" of matter, to the existence of which no limit can be set either in time or space, is the infinite and eternal substratum of all actual and possible existences, which is the doctrine of philosophical materialism, as I understand it, I have no objection to his holding that doctrine ; and I fail to comprehend how it can have the slightest influence upon any ethical or religious views he may please to hold.   If matter is the substratum of any phenomena of consciousness, animal or human, then it may possibly be the substratum of any other such phenomena ; if matter is imperishable, then it must be admitted to be possible that some of its combinations may be indefinitely enduring, just as our present so-called "elements" are probably only compounds which have been indissoluble, in our planet, for millions of years.   Moreover, the ultimate forms of existence which we distinguish in our little speck of the universe are, possibly, only two out of infinite varieties of existence, not only analogous to matter and analogous to mind, but of kinds which we are not competent so much as to conceive—in the midst of which, indeed, we might be set down, with no more notion of what was about us, than the worm in a flower-pot, on a London balcony, has of the life of the great city.

That which I do very strongly object to is the habit, which a great many non-philosophical materialists unfortunately fall into, of forgetting all these very obvious considerations.   They talk as if the proof that the "substance of matter" was the "substance" of all things cleared up all the mysteries of existence.   In point of fact, it leaves them exactly where they were.

The philosophical Materialist who takes the trouble to comprehend Berkeley finds that strict logic carries him no further than some such answer as this to the philosophical Idealist : Well, if I cannot show that you are wrong, you cannot show that I am ; if I should happen to be right, your proofs of the impossibility of knowing anything but states of consciousness would be as valid as they are now ; moreover, your religious and ethical difficulties are just as great as mine.   The speculative game is drawn—let us get to practical work.

NOTE B (p. 255).

I am afraid this paragraph is very faulty, and indeed misleading.

Scholastic "Realism" means the doctrine that generic conceptions have an objective existence apart from the human mind. Conceptualism asserts that they exist only in the mind; nominalism, that general terms are mere names indicative of the similarities of objective existences.

Locke's assertion that "motion and figure are really in the manna" is essentially a piece of realism in the scholastic sense. Berkeley would reply motion and figure are purely mental existences—abolish all minds, and what becomes of them? But that does not make him into a conceptualist, still less into a nominalist; and though he may have reached his ultimate position through conceptualism, his position is quite different.

Berkeley differs from all his predecessors in affirming that the only substantial existence is the hypothetical substratum of mind or "spirit"; and that the whole phenomenal world consists of nothing more than affections of human (and other?) spirits by the divine spirit. Pushed to its logical extreme, his system passes into pantheism pure and simple.

NOTE C (p. 273).

To any one who possesses the faculty of squinting I recommend the following experiment. Take two of the ordinary figures of a cube, drawn for the stereoscope, and place them some few inches apart on a screen or wall, the proper right hand figure being on the left and the proper left on the right; then squint so as to see the left hand figure with the right eye and the right with the left eye. After a little practice, there will suddenly appear, at the point of intersection of the lines prolonging the two optic axes, and apparently, suspended in the air, a figure of a cube. And this image of the cube is so real that a pencil held in the hand can be moved all round it, or driven through it.

ON SENSATION AND THE UNITY OF
STRUCTURE OF SENSIFEROUS ORGANS

[1879.]

THE maxim that metaphysical inquiries are barren
of result, and that the serious occupation of the
mind with them is a mere waste of time and
labour, finds much favour in the eyes of the
many persons who pride themselves on the
possession of sound common sense; and we
sometimes hear it enunciated by weighty authorities, as if its natural consequence, the
suppression of such studies, had the force of a
moral obligation.

In this case, however, as in some others, those
who lay down the law seem to forget that a wise
legislator will consider, not merely whether his
proposed enactment is desirable, but whether
obedience to it is possible. For, if the latter
question is answered negatively, the former is
surely hardly worth debate.

Here, in fact, lies the pith of the reply to those who would make metaphysics contraband of intellect. Whether it is desirable to place a prohibitory duty upon philosophical speculations or not, it is utterly impossible to prevent the importation of them into the mind. And it is not a little curious to observe that those who most loudly profess to abstain from such commodities are, all the while, unconscious consumers, on a great scale, of one or other of their multitudinous disguises or adulterations. With mouths full of the particular kind of heavily buttered toast which they affect, they inveigh against the eating of plain bread. In truth, the attempt to nourish the human intellect upon a diet which contains no metaphysics is about as hopeful as that of certain Eastern sages to nourish their bodies without destroying life. Everybody has heard the story of the pitiless microscopist, who ruined the peace of mind of one of these mild enthusiasts by showing him the animals moving in a drop of the water with which, in the innocency of his heart, he slaked his thirst; and the unsuspecting devotee of plain common sense may look for as unexpected a shock when the magnifier of severe logic reveals the germs, if not the full-grown shapes, of lively metaphysical postulates rampant amidst his most positive and matter-of-fact notions.

By way of escape from the metaphysical Will-o'-

the-wisps generated in the marshes of literature and theology, the serious student is sometimes bidden to betake himself to the solid ground of physical science. But the fish of immortal memory, who threw himself out of the frying-pan into the fire, was not more ill advised than the man who seeks sanctuary from philosophical persecution within the walls of the observatory or of the laboratory. It is said that "meta-physics" owe their name to the fact that, in Aristotle's works, questions of pure philosophy are dealt with immediately after those of physics. If so, the accident is happily symbolical of the essential relations of things; for metaphysical speculation follows as closely upon physical theory as black care upon the horseman.

One need but mention such fundamental, and indeed indispensable, conceptions of the natural philosopher as those of atoms and forces: or that of attraction considered as action at a distance; or that of potential energy; or the antinomies of a vacuum and a plenum; to call to mind the metaphysical background of physics and chemistry; while, in the biological sciences, the case is still worse. What is an individual among the lower plants and animals? Are genera and species realities or abstractions? Is there such a thing as vital force, or does the name denote a mere relic of metaphysical fetichism? Is the doctrine of final causes legitimate or illegitimate? These

are a few of the metaphysical topics which are
suggested by the most elementary study of
biological facts. But, more than this, it may be
truly said that the roots of every system of
philosophy lie deep among the facts of physiology.
No one can doubt that the organs and the
functions of sensation are as much a part of the
province of the physiologist, as are the organs and
functions of motion, or those of digestion; and yet
it is impossible to gain an acquaintance with even
the rudiments of the physiology of sensation
without being led straight to one of the most
fundamental of all metaphysical problems. In
fact, the sensory operations have been, from
time immemorial, the battle-ground of philosophers.

I have more than once taken occasion to point
out that we are indebted to Descartes, who happened to be a physiologist as well as a philosopher,
for the first distinct enunciation of the essential
elements of the true theory of sensation. In
later times, it is not to the works of the philosophers, if Hartley and James Mill are excepted,
but to those of the physiologists, that we must
turn for an adequate account of the sensory
process. Haller's luminous, though summary,
account of sensation in his admirable "Primæ
Lineæ," the first edition of which was printed in
1747, offers a striking contrast to the prolixity
and confusion of thought which pervade Reid's

"Inquiry," of seventeen years' later date.[1] Even Sir William Hamilton, learned historian and acute critic as he was, not only failed to apprehend the philosophical bearing of long-established physiological truths; but, when he affirmed that there is no reason to deny that the mind feels at the finger points, and none to assert that the brain is the sole organ of thought,[2] he showed that he had not apprehended the significance of the revolution commenced, two hundred years before his time, by Descartes, and effectively followed up by Haller, Hartley, and Bonnet, in the middle of the last century.

In truth, the theory of sensation, except in one

---

[1] In justice to Reid, however, it should be stated that the chapters on sensation in the *Essays on the Intellectual Powers* (1785) exhibit a great improvement. He is, in fact, in advance of his commentator, as the note to Essay II. chap. ii. p. 248 of Hamilton's edition shows.

[2] Haller, amplifying Descartes, writes in the *Primæ Lineæ*, CCCLXVI.—"Non est adeo obscurum sensum omnem oriri ab objecti sensibilis impressione in nervum quemcumque corporis humani, et eamdem per eum nervum ad cerebrum pervenientem tunc demum representari animæ, quando cerebrum adtigit. Ut etiam hoc falsum sit animam inproximo per sensoria nervorumque ramos sentire." . . . DLVII.—"Dum ergo sentimus quinque diversissima entia conjunguntur : corpus quod sentimus : organi sensorii adfectio ab eo corpore : cerebri adfectio a sensorii percussione nata : in anima nata mutatio : animæ denique conscientia et sensationis adperceptio." Nevertheless, Sir William Hamilton gravely informs his hearers :—"We have no more right to deny that the mind feels at the finger points, as consciousness assures us, than to assert that it thinks exclusively in the brain."—*Lecture on Metaphysics and Logic*, ii. p. 128. "We have no reason whatever to doubt the report of consciousness, that we actually perceive at the external point of sensation, and that we perceive the material reality."—*Ibid.* p. 129.

point, is, at the present moment, very much where Hartley, led by a hint of Sir Isaac Newton's, left it, when, a hundred and twenty years since, the "Observations on Man : his Frame, his Duty, and his Expectations," was laid before the world. The whole matter is put in a nutshell in the following passages of this notable book.

"External objects impressed upon the senses occasion, first on the nerves on which they are impressed, and then on the brain, vibrations of the small and, as we may say, infinitesimal medullary particles.

"These vibrations are motions backwards and forwards of the small particles ; of the same kind with the oscillations of pendulums and the tremblings of the particles of sounding bodies. They must be conceived to be exceedingly short and small, so as not to have the least efficacy to disturb or move the whole bodies of the nerves or brain." [1]

"The white medullary substance of the brain is also the immediate instrument by which ideas are presented to the mind ; or, in other words, whatever changes are made in this substance, corresponding changes are made in our ideas ; and vice versa." [2]

Hartley, like Haller, had no conception of the nature and functions of the grey matter of the brain. But, if for "white medullary substance," in the latter paragraph, we substitute "grey cellular substance," Hartley's propositions embody

---

[1] *Observations on Man,* vol. i. p. 11.
[2] *Ibid.* p. 8. The speculations of Bonnet are remarkably similar to those of Hartley ; and they appear to have originated independently, though the *Essai de Psychologie* (1754) is of five years' later date than the *Observations on Man* (1749).

the most probable conclusions which are to be drawn from the latest investigations of physiologists. In order to judge how completely this is the case, it will be well to study some simple case of sensation, and, following the example of Reid and of James Mill, we may begin with the sense of smell. Suppose that I become aware of a musky scent, to which the name of " muskiness " may be given. I call this an odour, and I class it along with the feelings of light, colours, sounds, tastes, and the like, among those phenomena which are known as sensations. To say that I am aware of this phenomenon, or that I have it, or that it exists, are simply different modes of affirming the same facts. If I am asked how I know that it exists, I can only reply that its existence and my knowledge of it are one and the same thing; in short, that my knowledge is immediate or intuitive, and, as such, is possessed of the highest conceivable degree of certainty.

The pure sensation of muskiness is almost sure to be followed by a mental state which is not a sensation, but a belief, that there is somewhere, close at hand, a something on which the existence of the sensation depends. It may be a musk-deer, or a musk-rat, or a musk-plant, or a grain of dry musk, or simply a scented handkerchief; but former experience leads us to believe that the sensation is due to the presence of one or other of these objects, and that it will vanish if the object is removed. In other words, there arises a belief in an external cause of the muskiness, which, in common language, is termed an odorous body.

But the manner in which this belief is usually put into words is strangely misleading. If we are dealing with a musk-plant, for example, we do not confine ourselves to a simple statement of that which we believe, and say that the musk-plant is the cause of the sensation called muskiness ; but we say that the plant has a musky smell, and we speak of the odour as a quality, or property, inherent in the plant. And the inevitable reaction of words upon thought has in this case become so complete, and has penetrated so deeply, that when an accurate statement of the case—namely, that muskiness, inasmuch as the term denotes nothing but a sensation, is a mental state, and has no existence except as a mental phenomenon—is first brought under the notice of common-sense folks, it is usually regarded by them as what they are pleased to call a mere metaphysical paradox and a patent example of useless subtlety. Yet the slightest reflection must suffice to convince any one possessed of sound reasoning faculties, that it is as absurd to suppose that muskiness is a quality inherent in one plant, as it would be to imagine that pain is a quality inherent in another, because we feel pain when a thorn pricks the finger.

Even the common-sense philosopher, *par excel-*

lence, says of smell: " It appears to be a simple and original affection or feeling of the mind, altogether inexplicable and unaccountable. It is indeed impossible that it can be in any body: it is a sensation, and a sensation can only be in a sentient thing." [1]

That which is true of muskiness is true of every other odour. Lavender-smell, clove-smell, garlic-smell, are, like "muskiness," names of states of consciousness, and have no existence except as such. But, in ordinary language, we speak of all these odours as if they were independent entities residing in lavender, cloves, and garlic; and it is not without a certain struggle that the false metaphysic of so-called common sense, thus ingrained in us, is expelled.

For the present purpose, it is unnecessary to inquire into the origin of our belief in external bodies, or into that of the notion of causation. Assuming the existence of an external world, there is no difficulty in obtaining experimental proof that, as a general rule, olfactory sensations,

---

[1] *An Inquiry into the Human Mind on the Principles of Common Sense,* chap. ii. § 2. Reid affirms that "it is genius, and not the want of it, that adulterates philosophy, and fills it with error and false theory;" and no doubt his own lucubrations are free from the smallest taint of the impurity to which he objects. But, for want of something more than that sort of "common sense," which is very common and a little dull, the contemner of genius did not notice that the admission here made knocks so big a hole in the bottom of "common sense philosophy," that nothing can save it from foundering in the dreaded abyss of Idealism.

are caused by odorous bodies; and we may pass on to the next step of the inquiry—namely, how the odorous body produces the effect attributed to it.

The first point to be noted here is another fact revealed by experience; that the appearance of the sensation is governed, not only by the presence of the odorous substance, but by the condition of a certain part of our corporeal structure, the nose. If the nostrils are closed, the presence of the odorous substance does not give rise to the sensation; while, when they are open, the sensation is intensified by the approximation of the odorous substance to them, and by snuffing up the adjacent air in such a manner as to draw it into the nose. On the other hand, looking at an odorous substance, or rubbing it on the skin, or holding it to the ear, does not awaken the sensation. Thus, it can be readily established by experiment that the perviousness of the nasal passages is, in some way, essential to the sensory function; in fact, that the organ of that function is lodged somewhere in the nasal passages. And, since odorous bodies give rise to their effects at considerable distances, the suggestion is obvious that something must pass from them into the sense organ. What is this "something," which plays the part of an intermediary between the odorous body and the sensory organ?

The oldest speculation about the matter dates

298 SENSATION AND THE SENSIFEROUS ORGANS

back to the Epicurean School and Democritus, and it is to be found fully stated in the fourth book of Lucretius. It comes to this: that the surfaces of bodies are constantly throwing off excessively attenuated films of their own substance: and that these films, reaching the mind, excite the appropriate sensations in it.

Aristotle did not admit the existence of any such material films, but conceived that it was the form of the substance, and not its matter, which affected sense, as a seal impresses wax, without losing anything in the process. While many, if not the majority, of the Schoolmen took up an intermediate position and supposed that a something, which was not exactly either material or immaterial, and which they called an "intentional species," effected the needful communication between the bodily cause of sensation and the mind.

But all these notions, whatever may be said for or against them in general, are fundamentally defective, by reason of an oversight which was inevitable, in the state of knowledge at the time in which they were promulgated. What the older philosophers did not know, and could not know, before the anatomist and the physiologist had done their work, is that, between the external object and that mind in which they supposed the sensation to inhere, there lies a physical obstacle. The sense organ is not a mere passage by which

the "tenuia simulacra rerum," or the "intentional
species" cast off by objects, or the "forms" of
sensible things, pass straight to the mind ; on the
contrary, it stands as a firm and impervious
barrier, through which no material particle of
the world without can make its way to the world
within.

Let us consider the olfactory sense organ more
nearly. Each of the nostrils leads into a passage
completely separated from the other by a partition, and these two passages place the nostrils in
free communication with the back of the throat,
so that they freely transmit the air passing to the
lungs when the mouth is shut, as in ordinary
breathing. The floor of each passage is flat, but
its roof is a high arch, the crown of which is
seated between the orbital cavities of the skull,
which serve for the lodgment and protection of
the eyes; and it therefore lies behind the apparent limits of that feature which, in ordinary
language, is called the nose. From the side walls
of the upper and back part of these arched chambers, certain delicate plates of bone project, and
these, as well as a considerable part of the
partition between the two chambers, are covered
by a fine, soft, moist membrane. It is to this
"Schneiderian," or olfactory, membrane that
odorous bodies must obtain direct access, if they
are to give rise to their appropriate sensations ;
and it is upon the relatively large surface, which

the olfactory membrane offers, that we must seek
for the seat of the organ of the olfactory sense.
The only essential part of that organ consists of a
multitude of minute rod-like bodies, set perpen-
dicularly to the surface of the membrane, and
forming a part of the cellular coat, or epithelium,
which covers the olfactory membrane, as the
epidermis covers the skin. In the case of the
olfactory sense, there can be no doubt that the
Democritic hypothesis, at any rate for such
odorous substances as musk, has a good foundation. Infinitesimal particles of musk fly off from
the surface of the odorous body; these, becoming
diffused through the air, are carried into the nasal
passages, and thence into the olfactory chambers,
where they come into contact with the filamen-
tous extremities of the delicate olfactory
epithelium.

But this is not all. The "mind" is not, so to
speak, upon the other side of the epithelium. On
the contrary, the inner ends of the olfactory cells
are connected with nerve fibres, and these nerve
fibres, passing into the cavity of the skull, at
length end in a part of the brain, the olfactory
sensorium. It is certain that the integrity of
each, and the physical inter-connection of all these
three structures, the epithelium of the sensory
organ, the nerve fibres, and the sensorium, are
essential conditions of ordinary sensation. That
is to say, the air in the olfactory chambers may be

charged with particles of musk; but, if either the epithelium, or the nerve fibres, or the sensorium is injured, or if they are physically disconnected from one another, sensation will not arise. Moreover, the epithelium may be said to be receptive, the nerve fibres transmissive, and the sensorium sensifacient. For, in the act of smelling, the particles of the odorous substance produce a molecular change (which Hartley was in all probability right in terming a vibration) in the epithelium, and this change being transmitted to the nerve fibres, passes along them with a measurable velocity, and, finally reaching the sensorium, is immediately followed by the sensation.

Thus, modern investigation supplies a representative of the Epicurean "simulacra" in the volatile particles of the musk; but it also gives us the stamp of the particles on the olfactory epithelium, without any transmission of matter, as the equivalent of the Aristotelian "form"; while, finally, the modes of motion of the molecules of the olfactory cells, of the nerve, and of the cerebral sensorium, which are Hartley's vibrations, may stand very well for a double of the "intentional species" of the Schoolmen. And this last remark is not intended merely to suggest a fanciful parallel; for, if the cause of the sensation is, as analogy suggests, to be sought in the mode of motion of the object of sense, then it is quite possible that the particular mode of motion of the object is reproduced in the sensorium; exactly as the diaphragm of a telephone reproduces the mode of motion taken up at its receiving end. In other words, the secondary "intentional species" may be, as the Schoolmen thought the primary one was, the last link between matter and mind.

None the less, however, does it remain true that no similarity exists, nor indeed is conceivable, between the cause of the sensation and the sensation. Attend as closely to the sensations of muskiness, or any other odour, as we will, no trace of extension, resistance, or motion is discernible in them. They have no attribute in common with those which we ascribe to matter; they are, in the strictest sense of the words, immaterial entities.

Thus, the most elementary study of sensation justifies Descartes' position, that we know more of mind than we do of body; that the immaterial world is a firmer reality than the material. For the sensation "muskiness" is known immediately. So long as it persists, it is a part of what we call our thinking selves, and its existence lies beyond the possibility of doubt. The knowledge of an objective or material cause of the sensation, on the other hand, is mediate; it is a belief as contradistinguished from an intuition; and it is a belief which, in any given instance of sensation, may, by possibility, be devoid of foundation. For odours, like other sensations, may arise from the occurrence of the appropriate molecular changes in the nerve or in the sensorium, by the operation of a cause distinct from the affection of the sense organ by an odorous body. Such "subjective" sensations are as real existences as any others, and as distinctly suggest an external odorous object as their cause; but the belief thus generated is a delusion. And, if beliefs are properly termed "testimonies of consciousness," then undoubtedly the testimony of consciousness may be, and often is, untrustworthy.

Another very important consideration arises out of the facts as they are now known. That which, in the absence of a knowledge of the physiology of sensation, we call the cause of the smell, and term the odorous object, is only such, mediately, by reason of its emitting particles which give rise to a mode of motion in the sense organ. The sense organ, again, is only a mediate cause by reason of its producing a molecular change in the nerve fibre; while this last change is also only a mediate cause of sensation, depending, as it does, upon the change which it excites in the sensorium.

The sense organ, the nerve, and the sensorium, taken together, constitute the sensiferous apparatus. They make up the thickness of the wall between the mind, as represented by the sensation "muskiness," and the object, as represented by the particle of musk in contact with the olfactory epithelium.

It will be observed that the sensiferous wall and the external world are of the same nature; whatever it is that constitutes them both is expressible in terms of matter and motion. Whatever changes take place in the sensiferous apparatus are continuous with, and similar to, those which take place in the external world.[1] But, with the sensorium, matter and motion come to an end; while phenomena of another order, or immaterial states of consciousness, make their appearance. How is the relation between the material and the immaterial phenomena to be conceived? This is

[1] The following diagrammatic scheme may help to elucidate the theory of sensation :—

Mediate Knowledge			Immediate Knowledge
	Sensiferous Apparatus		
Objects of Sense	Receptive. Transmissive. Sensificatory (Sense Organ) (Nerve) (Sensorium)		Sensations and other States of Consciousness
Hypothetical Substance of Matter			Hypothetical Substance of Mind
Physical World			Mental World
Not Self		Self	
Non-Ego or Object			Ego or Subject

Immediate knowledge is confined to states of consciousness, or, in other words, to the phenomena of mind. Knowledge of the physical world, or of one's own body and of objects external to it, is a system of beliefs or judgments based on the sensations. The term "self" is applied not only to the series of mental phenomena which constitute the ego, but to the fragment of the physical world which is their constant concomitant. The corporeal self, therefore, is part of the non-ego ; and is objective in relation to the ego as subject.

the metaphysical problem of problems, and the solutions which have been suggested have been made the corner-stones of systems of philosophy. Three mutually irreconcilable readings of the riddle have been offered.

The first is, that an immaterial substance of mind exists; and that it is affected by the mode of motion of the sensorium, in such a way as to give rise to the sensation.

The second is, that the sensation is a direct effect of the mode of motion of the sensorium, brought about without the intervention of any substance of mind.

The third is, that the sensation is, neither directly nor indirectly, an effect of the mode of motion of the sensorium, but that it has an independent cause. Properly speaking, therefore, it is not an effect of the motion of the sensorium, but a concomitant of it.

As none of these hypotheses is capable of even an approximation to demonstration, it is almost needless to remark that they have been severally held with tenacity and advocated with passion. I do not think it can be said of any of the three that it is inconceivable, or that it can be assumed on à priori grounds to be impossible.

Consider the first, for example; an immaterial substance is perfectly conceivable. In fact, it is obvious that, if we possessed no sensations but those of smell and hearing, we should be unable to conceive a material substance. We might have a conception of time, but could have none of extension, or of resistance, or of motion. And without the three latter conceptions no idea of matter could be formed. Our whole knowledge would be limited to that of a shifting succession of immaterial phenomena. But if an immaterial substance may exist, it may have any conceivable properties; and sensation may be one of them. All these propositions may be affirmed with complete dialectic safety, inasmuch as they cannot possibly be disproved; but neither can a particle of demonstrative evidence be offered in favour of the existence of an immaterial substance.

As regards the second hypothesis, it certainly is not inconceivable, and therefore it may be true that sensation is the direct effect of certain kinds of bodily motion. It is just as easy to suppose this as to suppose, on the former hypothesis, that bodily motion affects an immaterial substance. But neither is it susceptible of proof.

And, as to the third hypothesis, since the logic of induction is in no case competent to prove that events apparently standing in the relation of cause and effect may not both be effects of a common cause—that also is as safe from refutation, if as incapable of demonstration, as the other two.

In my own opinion, neither of these speculations can be regarded seriously as anything but a more or less convenient working hypothesis. But, if I must choose among them, I take the "law of parcimony" for my guide, and select the simplest —namely, that the sensation is the direct effect of the mode of motion of the sensorium. It may justly be said that this is not the slightest explanation of sensation; but then am I really any the wiser, if I say that a sensation is an activity (of which I know nothing) of a substance of mind (of which also I know nothing) ? Or, if I say that the Deity causes the sensation to arise in my mind immediately after he has caused the particles of the sensorium to move in a certain way, is anything gained ? In truth, a sensation, as we have already seen, is an intuition—a part of immediate knowledge. As such, it is an ultimate fact and inexplicable; and all that we can hope to find out about it, and that indeed is worth finding out, is its relation to other natural facts. That relation appears to me to be sufficiently expressed, for all practical purposes, by saying that sensation is the invariable consequent of certain changes in the sensorium—or, in other words, that, so far as we know the change in the sensorium is the cause of the sensation.

I permit myself to imagine that the untutored, if noble, savage of "common sense" who has been misled into reading thus far, by the hope of getting positive solid information about sensation, giving way to not unnatural irritation, may here interpellate : " The upshot of all this long disquisition is that we are profoundly ignorant. We knew that to begin with, and you have merely furnished another example of the emptiness and uselessness of metaphysics." But I venture to reply, Pardon me, you were ignorant, but you did not know it. On the contrary, you thought you knew a great deal, and were quite satisfied with the particularly absurd metaphysical notions which you were pleased to call the teachings of common sense. You thought that your sensations were properties of external things, and had an existence outside of yourself. You thought that you knew more about material than you do about immaterial existences. And if, as a wise man has assured us, the knowledge of what we don't know is the next best thing to the knowledge of what we do know, this brief excursion into the province of philosophy has been highly profitable.

Of all the dangerous mental habits, that which schoolboys call " cocksureness " is probably the most perilous; and the inestimable value of metaphysical discipline is that it furnishes an effectual counterpoise to this evil proclivity. Whoso has mastered the elements of philosophy knows that the attribute of unquestionable certainty appertains only to the existence of a state of consciousness so long as it exists; all other beliefs are mere probabilities of a higher or lower order. Sound metaphysic is an amulet which renders its possessor proof alike against the poison of superstition and the counter-poison of shallow negation; by showing that the affirmations of the former and the denials of the latter alike deal with matters about which, for lack of evidence, nothing can be either affirmed or denied.

I have dwelt at length upon the nature and origin of our sensations of smell, on account of the comparative freedom of the olfactory sense from the complications which are met with in most of the other senses.

Sensations of taste, however, are generated in almost as simple a fashion as those of smell. In this case, the sense organ is the epithelium which covers the tongue and the palate: and which sometimes, becoming modified, gives rise to peculiar organs termed " gustatory bulbs," in which the epithelial cells elongate and assume a somewhat rodlike form. Nerve fibres connect the sensory organ with the sensorium, and tastes or flavours are states of consciousness caused by the change of molecular state of the latter. In the case of the sense of touch there is often no sense organ distinct from the general epidermis. But many fishes and amphibia exhibit local modifications of the epidermic cells which are, sometimes, extraordinarily like the gustatory bulbs; more commonly, both in lower and higher animals, the effect of the contact of external bodies is intensified by the development of hair-like filaments, or of true hairs, the bases of which are in immediate relation with the ends of the sensory nerves. Every one must have noticed the extreme delicacy of the sensations produced by the contact of bodies with the ends of the hairs of the head; and the " whiskers " of cats owe their functional importance to the abundant supply of nerves to the follicles in which their bases are lodged. What part, if any, the so-called "tactile corpuscles," " end bulbs," and " Pacinian bodies," play in the mechanism of touch is unknown. If they are sense organs, they are exceptional in character, in so far as they do not appear to be modifications of the epidermis. Nothing is known respecting the organs of those sensations of resistance which are grouped under the head of the muscular sense; nor of the sensations of warmth and cold; nor of that very singular sensation which we call tickling.

In the case of heat and cold, the organism not only becomes affected by external bodies, far more remote than those which affect the sense of smell; but the Democritic hypothesis is obviously no longer permissible. When the direct rays of the sun fall upon the skin, the sensation of heat is certainly not caused by "attenuated films " thrown off from that luminary, but is due to a mode of motion which is transmitted to us. In Aristotelian phrase, it is the form without the matter of the sun which stamps the sense organ; and this, translated into modern language, means nearly the same thing as Hartley's vibrations. Thus we are prepared for what happens in the case of the auditory and the visual senses. For neither the ear, nor the eye, receives anything but the impulses or vibrations originated by sonorous or luminous bodies. Nevertheless, the receptive apparatus still consists of specially modified epithelial cells. In the labyrinth, or essential part of the ear of the higher animals, the free ends of these cells terminate in excessively delicate hair-like filaments; while, in the lower forms of auditory organ, its free surface is beset with delicate hairs like those of the surface of the body, and the transmissive nerves are connected with the bases of these hairs. Thus there is an insensible gradation in the forms of the receptive apparatus, from the organ of touch, on the one hand, to those of taste and smell; and, on the other hand, to that of hearing.

Even in the case of the most refined of all the sense organs, that of vision, the receptive apparatus departs but little from the general type. The only essential constituent of the visual sense organ is the retina, which forms so small a part of the eyes of the higher animals; and the simplest eyes are nothing but portions of the integument, in which the cells of the epidermis have become converted into glassy rod-like retinal corpuscles. The outer ends of these are turned towards the light; their sides are more or less extensively coated with a dark pigment, and their inner ends are connected with the transmissive nerve fibres. The light, impinging on these visual rods, produces a change in them which is communicated to the nerve fibres, and, being transmitted to the sensorium, gives rise to the sensation—if indeed all animals which possess eyes are endowed with what we understand as sensation.

In the higher animals, a complicated apparatus of lenses, arranged on the principle of a camera obscura, serves at once to concentrate and to individualise the pencils of light proceeding from external bodies. But the essential part of the organ of vision is still a layer of cells, which have the form of rods with truncated or conical ends. By what seems a strange anomaly, however, the glassy ends of these are turned not towards, but away from, the light: and the latter has to traverse the layer of nervous tissues with which their outer ends are connected, before it can affect them. Moreover, the rods and cones of the vertebrate retina are so deeply seated, and in many respects so peculiar in character, that it appears impossible, at first sight, that they can have anything to do with that epidermis of which gustatory and tactile and, at any rate, the lower forms of auditory and visual, organs are obvious modifications.

Whatever be the apparent diversities among the sensiferous apparatuses, however, they share certain common characters. Each consists of a receptive, a transmissive, and a sensificatory portion. The essential part of the first is an epithelium, of the second, nerve fibres, of the third, a part of the brain; the sensation is always the consequence of the mode of motion excited in the receptive, and sent along the transmissive, to the sensificatory part of the sensiferous apparatus. And, in all the senses, there is no likeness whatever between the object of sense, which is matter in motion, and the sensation, which is an immaterial phenomenon.

On the hypothesis which appears to me to be the most convenient, sensation is a product of the sensiferous apparatus caused by certain modes of motion which are set up in it by impulses from without. The sensiferous apparatuses are, as it were, factories, all of which at the one end receive raw materials of a similar kind—namely, modes of motion—while, at the other, each turns out a special product, the feeling which constitutes the kind of sensation characteristic of it.

Or, to make use of a closer comparison, each sensiferous apparatus is comparable to a musical-box wound up; with as many tunes as there are separate sensations. The object of a simple sensation is the agent which presses down the stop of one of these tunes, and the more feeble the agent, the more delicate must be the mobility of the stop.[1]

But, if this be true, if the recipient part of the sensiferous apparatus is in all cases, merely a mechanism affected by coarser or finer kinds of material motion, we might expect to find that all sense organs are fundamentally alike, and result from the modification of the same morphological elements. And this is exactly what does result from all recent histological and embryological investigations.

It has been seen that the receptive part of the olfactory apparatus is a slightly modified epithelium, which lines an olfactory chamber deeply seated between the orbits in adult human beings. But, if we trace back the nasal chambers to their origin in the embryo, we find, that, to begin with, they are mere depressions of the skin of the forepart of the head, lined by a continuation of the general epidermis. These depressions become pits, and the pits, by the growth of the adjacent parts, gradually acquire the position which they finally occupy. The olfactory organ, therefore, is a specially modified part of the general integument.

---

[1] "Chaque fibre est une espèce de touche ou de marteau destiné à rendre un certain ton."—Bonnet, *Essai de Psychologie*, chap. iv.

The human ear would seem to present greater difficulties. For the essential part of the sense organ, in this case, is the membranous labyrinth, a bag of complicated form, which lies buried in the depths of the floor of the skull, and is surrounded by dense and solid bone. Here, however, recourse to the study of development readily unravels the mystery. Shortly after the time when the olfactory organ appears, as a depression of the skin on the side of the fore part of the head, the auditory organ appears as a similar depression on the side of its back part. The depression, rapidly deepening, becomes a small pouch; and then, the communication with the exterior becoming shut off, the pouch is converted into a closed bag, the epithelial lining of which is a part of the general epidermis segregated from the rest. The adjacent tissues, changing first into cartilage and then into bone, enclose the auditory sac in a strong case, in which it undergoes its further metamorphoses; while the drum, the ear bones, and the external ear, are superadded by no less extraordinary modifications of the adjacent parts. Still more marvellous is the history of the development of the organ of vision. In the place of the eye, as in that of the nose and that of the ear, the young embryo presents a depression of the general integument; but, in man and the higher animals, this does not give rise to the proper sensory organ, but only to part of the accessory structures concerned in vision. In fact, the depression, deepening and becoming converted into a shut sac, produces only the cornea, the aqueous humour, and the crystalline lens of the perfect eye.

The retina is added to this by the outgrowth of the wall of a portion of the brain into a sort of bag, or sac, with a narrow neck, the convex bottom of which is turned outwards, or towards the crystalline lens. As the development of the eye proceeds, the convex bottom of the bag becomes pushed in, so that it gradually obliterates the cavity of the sac, the previously convex wall of which becomes deeply concave. The sac of the brain is now like a double nightcap ready for the head, but the place which the head would occupy is taken by the vitreous humour, while the layer of nightcap next it becomes the retina. The cells of this layer which lie farthest from the vitreous humour, or, in other words, bound the original cavity of the sac, are metamorphosed into the rods and cones. Suppose now that the sac of the brain could be brought back to its original form ; then the rods and cones would form part of the lining of a side pouch of the brain. But one of the most wonderful revelations of embryology is the proof of the fact that the brain itself is, at its first beginning, merely an infolding of the epidermic layer of the general integument. Hence it follows that the rods and cones of the vertebrate eye are modified epidermic cells, as much as the crystalline cones of the insect or crustacean eye are ; and that the inversion of the position of the former in relation to light arises simply from the roundabout way in which the vertebrate retina is developed.

Thus all the higher sense organs start from one foundation, and the receptive epithelium of the eye, or of the ear, is as much modified epidermis as is that of the nose. The structural unity of the sense organs is the morphological parallel to their identity of physiological function, which, as we have seen, is to be impressed by certain modes of motion; and they are fine or coarse, in proportion to the delicacy or the strength of the impulses by which they are to be affected.

In ultimate analysis, then, it appears that a sensation is the equivalent in terms of consciousness for a mode of motion of the matter of the sensorium. But, if inquiry is pushed a stage farther, and the question is asked, What then do we know about matter and motion ? there is but one reply possible. All that we know about motion is that it is a name for certain changes in the relations of our visual, tactile, and muscular sensations; and all that we know about matter is that it is the hypothetical substance of physical phenomena— the assumption of the existence of which is as pure a piece of metaphysical speculation as is that of the existence of the substance of mind.

Our sensations, our pleasures, our pains, and the relations of these, make up the sum total of the elements of positive, unquestionable knowledge. We call a large section of these sensations and their relations matter and motion; the rest we term mind and thinking; and experience shows that there is a certain constant order of succession between some of the former and some of the latter.

This is all that just metaphysical criticism leaves of the idols set up by the spurious metaphysics of vulgar common sense. It is consistent either with pure Materialism, or with pure Idealism, but it is neither. For the Idealist, not content with declaring the truth that our knowledge is limited to facts of consciousness, affirms the wholly un-provable proposition that nothing exists beyond these and the substance of mind. And, on the other hand, the Materialist, holding by the truth that, for anything that appears to the contrary material phenomena are the causes of mental phenomena, asserts his unprovable dogma, that material phenomena and the substance of matter are the sole primary existences.

Strike out the propositions about which neither controversialist does or can know anything, and there is nothing left for them to quarrel about. Make a desert of the Unknowable, and the divine Astræa of philosophic peace will commence her blessed reign.

END OF VOL. VI

RICHARD CLAY AND SONS, LIMITED, LONDON AND BUNGAY.